Urban World/ Global City

Second edition

The last millennium marked a symbolic transition in the history of human settlement. Over half of the world's six billion people now live in towns and cities. The world is an urban place.

This book identifies and accounts for the characteristics of the contemporary city and of urban society. It analyses the distribution and growth of settlements and explores the social and behavioural characteristics of urban living. The latest theoretical and empirical developments and insights are synthesised and presented in an accessible and engaging way.

Emphasis throughout is placed upon the world scale with urban developments being seen as the geographical consequences of the evolution of capitalism. Individual chapters focus upon populations and places, growth and urbanisation, urban development as a global phenomenon, socio-economic consequences of global urban development, urban culture and global urban society, world cities and the urban future.

This second edition has been extensively updated and referenced. Each chapter includes learning objectives, annotated readings and topics for discussion. Well illustrated throughout, it will be essential reading for students of geography, sociology and development studies and all who seek an understanding of how the urban world has evolved and how it will change in the twenty-first century.

David Clark is Professor of Urban Geography at Coventry University.

Urban World / Global City

Second edition

David Clark

Routledge
Taylor & Francis Group

LONDON AND NEW YORK

First published 1996
by Routledge
11 New Fetter Lane, London EC4P 4EE

Simultaneously published in the USA and Canada
by Routledge
29 West 35th Street, New York, NY 10001

2nd edition 2003

Routledge is an imprint of the Taylor & Francis Group

Typeset in Times by Florence Production Ltd, Stoodleigh, Devon
Printed and bound in Great Britain by The Cromwell Press, Trowbridge, Wiltshire

British Library Cataloguing in Publication Data
A catalogue record for this book is available from the British Library

Library of Congress Cataloging in Publication Data
A catalog record for this book has been requested

ISBN 0–415–32097–6 (hbk)
ISBN 0–415–32098–4 (pbk)

Contents

Plates

⬭ Figures

⬤ Tables

Boxes

Preface

One hundred years ago, in 1899, Adna Ferrin Weber submitted a Ph.D. thesis to Columbia University on *The Growth of Cities in the Nineteenth Century*. This masterpiece of data collection, analysis and explanation was the first comprehensive attempt to document and understand urban phenomena at the global scale. It identified variations in levels of urban development around the world and explored in detail the social and economic differences between town and country. Weber's work was a seminal contribution to the understanding of cities and was responsible for establishing urban study as a central focus of inquiry in the newly emerging discipline of geography.

Although cities were commonplace in that part of North America in which Weber lived and worked, they were comparatively small and thinly scattered and together housed a minority of the population, a pattern that was repeated across much of north-western Europe. As places of residence they were different to rural settlements and their populations engaged in social and economic activities that were distinctively urban in character. Elsewhere in North America and the rest of the world, urban development was negligible and the population was overwhelmingly rural both in terms of location and ways of life. This pattern was slow to change. It is only in the last 30 years that it has become valid to talk about an urban world: a world in which urban places and urban living are the norm rather than the exception.

The contemporary urban world is a product of three principal developments. The first is the growth in the size, number and spread of settlements, so that there are now few regions that lack urban populations and places. The second is the increase in the proportion of the population that lives in urban places. The third is the transformation of society because so many people and such a large proportion of the world's population live in towns and cities and follow lifestyles which are urban in origin and character. These three developments are powerful, deep-seated and self-reinforcing. Together they have created an urban world that would be unrecognisable to analysts writing 100 years ago.

The urban world is, however, far from uniform. Urban development is changing the spatial organisation of economy and society, but at different rates in different places and with different outcomes. There are wide variations in the number and proportion of the population that live in urban settlements, the size and role of cities, and in the extent to which people live urban ways of life. The present pattern is best seen as a spectrum of development with urban predominating at one extreme, rural at the other, and a number of gradations in between. It is the remit of geographers to analyse and account for these contrasts and variations and to examine their outcomes and implications.

Urban growth, urbanisation and the spread of urbanism are seen in this text as products of the pronounced globalisation of economic and social activity that has occurred in recent years. They are viewed as consequences of a fundamental upward shift in the overall scale of geographical activity and organisation. Markets which were previously separate and localised have become merged and have been superseded by worldwide patterns of production and consumption coordinated by global institutions and organisations. Social patterns and relationships that were introspective and parochial have become open and global in extent. Globalisation is made possible by developments in transport, telecommunications and geopolitical convergence. It is grounded, it is argued, in a contemporary social and economic formation that is the most recent product of a capitalist mode of accumulation.

This book identifies, and seeks to account for, the characteristics of the contemporary urban world and global urban society. It traces the growth of towns and cities and the spread of urbanism, and attempts to explain the organisational characteristics of the global urban system. Emphasis throughout is placed upon the world scale, with variations in levels of urban development within nation states, which may be pronounced, being the focus of secondary attention. What appears between these covers is a broad overview and synthesis of global urban processes, patterns and products. What results, it is hoped, is a contribution to the understanding of how the urban world and global city are formed, function and are likely to evolve.

Thanks are due to members of the Geography staff at Coventry University, who provided much helpful comment and advice on the manuscript. I am especially grateful to Dr Hazel Barrett, who allowed me to use many of her excellent photographs, and to Shirley Addleton and Stuart Gill, who prepared the maps and diagrams.

Acknowledgements are also due to the countless undergraduate students, who unknowingly acted as sounding boards for some of the arguments and ideas. Any errors of omission, oversimplification or misinterpretation are mine.

David Clark
Ashow, 2003

Acknowledgements

The author and publishers would like to thank the following for granting permission to reproduce the plates stated. Dr Hazel Barrett: Plates 1.1, 1.2, 2.1, 2.2, 3.2, 4.1. 5.1, 5.2, 5.4, 5.5, 5.6, 5.7, 5.8. Photodisc Library: Plates 7.2, 7.3, 7.4, 7.5, 8.1.

1 Global patterns and perspectives

By the end of this chapter you should:

- be aware of the contribution of geography to the understanding of urban places;
- have a broad appreciation of the meaning and value of global perspectives in understanding the contemporary world;
- have a basic familiarity with the debates which surround the concept of globalisation.

Introduction

The recent millennium was a major watershed in the evolution of human settlement, for it marked the period when the location of the world's people became more urban than rural. It is likely that the figure of 50 per cent urban was achieved at some point between 1996 and 2001 but it is not possible to be exact because of variations among countries in the quality of their census data and in the ways in which urban areas are defined. Despite its geographical significance, this historical transition went largely unrecognised and unreported, so its profound symbolic importance was overlooked. More of the six billion inhabitants of the globe now live in towns and cities than in villages and hamlets. No longer are towns and cities exceptional settlement forms in predominantly rural societies. The world is an urban place.

Urban development on this scale is a remarkable geographical phenomenon. Far from being spread widely and thinly across the surface of the habitable earth, a population which is urban is one in which vast numbers of people are clustered together in very small areas. Whether through choice or compulsion, they live in close horizontal and vertical proximity and at very high densities. They seemingly prefer, or are forced to accept, concentration rather than dispersal. The benefits of access to services and other people, which are a consequence of closeness and agglomeration, apparently outweigh the disadvantages and drawbacks of crowding, congestion, noise and pollution. If the size of population is any guide, then living in an urban environment has greater appeal than residing in the countryside. The number and size of cities and the rate at which many of them are growing suggests that they are, and are likely to remain, highly attractive and acceptable forms of settlement to most people.

Cities are economic and social systems in space. They are a product of deep-seated and persistent processes which enable and encourage people to amass in large numbers in small areas. Surplus products, generated outside the city, provide the basic means of support, but viability and prosperity also depend on the existence of social arrangements and institutions through which cities are regulated and managed. So powerful and pervasive are the forces of urban formation and growth that they presently concentrate over three billion of the world's population in towns and cities. It is difficult adequately to convey an impression of the degree of clustering which this represents, as data on the combined area of all the world's urban places are not readily available. If, however, all the urban population lived at a density of 7,600 per sq. km, which is the average for the major cities of eastern Asia, then they could be accommodated on less than 1 per cent of the world's landmass, in an area roughly equivalent in size to Germany.

The urban world is distinctive in socio-economic as well as in spatial terms. Despite the infinite and intricate variations of tradition and culture that exist within and between nations, cities appear to have, and to be acquiring, more in common than they have differences. Urban places have many similarities of physical appearance, economic structure and social organisation and are beset by the same problems of employment, housing, health, transport and environmental quality. The elements in many urban skylines are the same, as commercial and residential areas are increasingly dominated by high-rise developments constructed in international styles. Streetscapes across the world are adjusting in the same way to accommodate the needs of the ubiquitous car, so cities are fast losing their individual layouts and architectural identities. Within buildings, workers do the same sorts of jobs, often on the same makes of computer or machine, and manufacture goods and services to the requirements of world markets dominated by a small number of global producers. Patterns of demand are converging as consumerism absorbs ever more of the world's population. There are few cities where McDonald's hamburgers (Plate 1.1), Fuji films, Microsoft's Office and

Plate 1.1 *Icons and images in the global city: a street in Jakarta, Indonesia*

Coca-Cola are not readily available and purchased in quantity. Some of these similarities are superficial and hide important underlying cultural differences, but the underlying trend is clear. There is increasing convergence among cities in physical and economic terms.

Irrespective of continent or country, many urban residents live their lives in broadly similar ways, with common concerns over home, children, school and work. Attitudes and expectations are shared as many aspire towards the lifestyles that are popularised and promoted by the mass media. Billions of people feast nightly on a diet of televised soap operas and international sporting events, with pop singers, film stars, sports personalities and media celebrities enjoying a worldwide recognition and following. Identification is reflected in fashions and accessories, with designer brands and labels such as Adidas, Tommy Hilfiger, Calvin Klein, Levi, Nike and Nokia commanding widespread following. Such interests, fads and tastes are increasingly independent of ethnicity, colour, class and creed. They draw together and fuse what geography and culture traditionally separate and divide. The contemporary urban world is more than a motley assemblage of diverse settlements. Many observers argue that it is slowly becoming a unitary and uniform place, a global city in which most of its inhabitants are imbued with a similar set of all-encompassing urban attitudes and values and follow common modes of behaviour.

Although towns and cities have existed for over eight millennia, the wholesale transition to urban location and urban living is very recent in origin. Many highly successful urban civilisations existed in the past, but their impacts were both limited and localised. In 1700 fewer than 2 per cent of the world's population lived in urban places and these were concentrated in a small number of city states. Major and rapid changes began in Britain in the late eighteenth century in response to the locational dictates of industrial capitalism. They subsequently spread to north-western Europe and north-eastern USA so that, by the beginning of the twentieth century, about 15 per cent of the world's population was living in urban places. Urban development as a global phenomenon is, however, essentially a feature of the last half of the twentieth century, indeed of its last three decades. Large parts of the world were effectively untouched by urban development and urban influences until 1970. They are occurring today because of massive changes in the distribution of population in countries which, until recently, were substantially and profoundly rural. Contemporary processes of urban development affect vast numbers of people across the globe. Although it is taking place at a local level, the present switch from rural to urban constitutes the largest shift in the location of population ever recorded.

Historical milestones are occasions for reflection and speculation; for looking back and to the future. They are a time for assessing what has been achieved and what opportunities, obstacles and likely outcomes lie ahead. The emergence of an urban world and the prospects for a global city pose important questions concerning the nature and consequences of the urban pattern and experience. They focus attention upon the reasons how and why cities exist, the ways in which they grow and their impact upon society. They raise issues concerning spatial and temporal variations in levels and rates of urban development and the implications of future urban change. Serious reservations surround the environmental and social sustainability of urban populations as the number and size of cities continue to increase. Such concerns challenge analysts to develop appropriate philosophies and methodologies by means of which the urban world can be conceptualised, explained and understood.

Studying the urban world

In view of this ambitious research agenda, it is not surprising to find that the task of analysis and explanation occupies an army of specialists drawn from a wide range of fields in the social and environmental sciences. No one discipline can claim to monopolise the study of the city, since urban questions and problems cut across many of the traditional divisions of academic inquiry. Equally, no single methodology predominates in urban analysis, for the complexities of urban life necessitate the adoption of a wide variety of approaches. It is in the interdisciplinary nature of urban issues that the city poses the greatest intellectual challenge to the analyst. Progress in urban study requires the fusion of insights derived from a number of subject areas, each of which approaches the analysis of urban settlements in its own distinctive way.

Geographers are prominent among the researchers who set out to analyse and explain the urban world and the global city. In focusing upon location they seek to add a spatial dimension to the understanding and interpretation of global urban phenomena (Box 1.1). Geographers

Box 1.1 The nature of geographical study

Geography is the academic discipline that explores the relationship between the Earth and its peoples through the study of place, space and the environment.

The study of place seeks to describe, explain and understand the location of the human and physical features of the Earth and the processes, systems and interrelationships that create or influence those features.

The study of space seeks to explore the relationships between places and patterns of activity from the use people make of the physical settings in which they live and work.

The study of environment embraces both its human and physical dimensions. It addresses the resources that the Earth provides, the impact upon those resources of human activities, and the wider economic, political and cultural consequences of the interrelationship between the two.

The spatial perspective upon phenomena that is adopted in geography is distinctive, even though geographers may look at the same phenomenon as other specialists. No other discipline has location and distribution as its major focus of study.

Source: DES and the Welsh Office (1990: 6).

are concerned to identify and account for the distribution and growth of towns and cities and the spatial similarities and contrasts that exist within and between them. They focus on both the contemporary urban pattern and the ways in which the distribution and internal arrangement of settlements have changed over time. They look at the ways in which cities are represented as places through appearances and images, how people identify with them and the effects upon ways of life and behaviour. Emphasis in urban geography is directed towards the understanding of those social and economic processes that determine the existence, evolution and functional organisation of urban places and the characteristics of urban society. In this way, geographical analysis both supplements and complements the insights provided by allied disciplines in the social sciences that recognise the urban world as a distinctive focus of study.

A wide range of approaches and methods is used by geographers in urban study (Box 1.2). Simple mapping of distributions is the starting point for most geographical work, since it identifies basic patterns and draws attention to possible causal relationships. It is especially appropriate at the global scale, where it is important to start with an overall picture and where variations, and hence the implications of urban development, are both complex and pronounced. Mapping, however, leads to little more than low-level description and is undermined by the highly variable quality of global urban data (see Appendix). Explanation is facilitated if attention is concentrated upon causal actions and mechanisms. Such relationships are products of tradition, culture and politics and are deeply embedded in the underlying strata that give societies and economies their form. A key task for geographers is to investigate and to understand the structural relationships that give rise to processes that in turn are responsible for creating observed urban patterns.

Although an urban specialism is long established in geography, the adoption of a world perspective on cities and urban society is a recent development. It was foreshadowed nearly 100 years ago in the formative work of Adna Ferin Weber (1899) on *The Growth of Cities in the Nineteenth Century*, although the urban 'world' that he analysed consisted of only around 50 countries in which there was significant urban development. This major empirical study was important because it showed how much earlier and further advanced were England, Wales and Scotland as urban societies and how little urban development there was outside north-western Europe and the eastern seaboard of the USA. The world at the time was very much a rural place in which

Box 1.2 Philosophical perspectives in urban geographical study

Geographers approach the study of urban worlds and global cities from a number of philosophical perspectives that contribute to explanation and understanding in different ways. Several are reflected in different sections of this book:

Positivism is the principle that underlies 'scientific' human geography. It assumes that there is regularity and uniformity in the distribution and characteristics of cities that can be observed, analysed and explained by an independent analyst. Positivist approaches seek to develop general rather than unique explanations. These are typically advanced in the form of models, theories and hypotheses of urban location. A major criticism is that observations are inherently subjective because analysts cannot stand outside society and distance themselves from it.

Behavioural approaches seek to explain urban patterns by focusing upon processes of decision-making. The emphasis is upon how people perceive the world and so behave towards it. The approach is of greatest value in micro-scale urban studies of, for example, shopping and migration. Perception, behaviour and decision-making are difficult to measure and to observe objectively. Individual differences tend to become blurred and so of limited importance in studies at aggregate scales.

Structuralism involves a belief that geographical patterns are grounded in underlying formations such as capitalism, rather than in superficial concepts like price, rent and profit. They may be products of power relationships that are studied by adherents of a *political economy* approach. Links between underlying structure and surface pattern, however, are difficult to disentangle, as they operate indirectly. Another criticism is that analysts following this approach are themselves products of the underlying structure and can only ever view patterns from the inside.

Proponents of *postmodern* approaches reject the idea of grand theory and emphasise instead the importance of differences, identities and representations. These may be reflected in buildings, iconography and signs. The prime concern is to try to understand the urban world from the multiple viewpoints of diverse individuals and groups. It is difficult, however, to capture the full range of differences in diverse populations. A further problem is to produce insights of general value and relevance out of understandings of the unique.

the development and distribution of towns and cities was limited and urban influences were restricted and localised.

Within urbanised countries, statistics on employment patterns, family structures and demography pointed to the existence of pronounced urban–rural contrasts. Cities were places with particular socio-economic characters that sustained and perpetuated distinctive patterns of social and economic behaviour. They were places in which urban ways of life evolved and were spread into the surrounding rural areas. As well as identifying the salient characteristics of the urban world, Weber's analysis was of considerable significance in a technical sense, since it drew attention to the many problems of data availability, quality and comparability that so bedevil urban analysis and understanding at the global scale.

The principal focus of subsequent work in urban geography was on systematic themes rather than overall global patterns. Important contributions were made to the understanding of urban lifestyles (Wirth, 1938), city size distributions (Zipf, 1949; Berry, 1961), urbanisation (Davis, 1965) and the colonial city (McGee, 1967), establishing traditions of research in these areas, but these studies directed attention towards constituent parts and individual phenomena rather than the urban world as a whole. Nowhere is this more evident than in the wealth of literature on urban growth and urbanisation as overviewed and synthesised by Pacione (2001). This invariably concentrates upon either the developing or the developed world; or else adopts a country by country or region by region approach. Such work has generated some powerful insights into the underlying patterns and processes, but the regional and national focus diverts attention away from wider similarities, possible common causation and the relevance of general theory.

A global economy?

The global view involves a focus upon scale and process, since it is concerned with worldwide patterns and the mechanisms which create them. There is nothing new in looking at the world as a whole, but it is only very recently that analysts have suggested that economic and social relationships at this level can be attributed to a small number of powerful forces that operate globally. This approach derives much of its contemporary popularity from the work of Immanuel Wallerstein on the nature and extent of the links between societies today and in the past. For Wallerstein (1979, 1980, 1989), most of the discrete localised economies that once existed are now merged and amalgamated into a single, integrated, world economy. This is global in organisation and reach and incorporates and encompasses the majority of the world's nation states and territories. Few regions, he argues, lie outside its limits and are untouched by its influence.

The world economy is capitalist in formation in that it is based upon principles of private rather than state ownership of the means of production and seeks to generate profits through the manipulation of land, labour, finance and entrepreneurship. It is primarily concerned with making and providing goods and services, be they computer equipment, cars, televisions, food, textiles, weapons or media products, for global consumption. Patterns of supply and demand, circulation and exchange, and marketing and advertising (Plate 1.2), together with all the economic and social structures that make them possible, are worldwide rather than local in scale. Social and economic outcomes in the form of opportunity, advantage, injustice and poverty are similarly global in extent and implication.

The world economy is distinguished by the ways in which it is organised and operates. Structure and function are the key defining features, rather than the worldwide scale of supply and demand, and of production and consumption. It is dominated by powerful transnational corporations (TNCs) and is regulated, through global institutions, by international agreements and conventions. Transnational corporations are large, complex companies that make and sell many products in many countries around the world. They dominate and control production and

Plate 1.2 *Promoting global consumption: advertising American cigarettes in Zambia*

consumption in key economic sectors. They have a disproportionate influence over supplies of raw materials and manufacturing capacity, and determine and direct patterns of spending through advertising and promotional activities. Transnational corporations are supported by banking and investment houses that manage and manipulate global finance, and by a range of organisations that provide producer services in the form of management consultancy and legal, personnel and marketing advice on an international basis. The organisation of the world economy is made possible by, and is maintained through, an international division of labour in which the tasks that people perform, their working conditions and their rates of pay are determined by the requirements of global capitalism. It is regulated by agreements and conventions and is policed by organisations such as the World Bank and the General Agreement on Tariffs and Trade, which define rules of behaviour and accountability.

Although it functions as an integrated whole, there are important structural variations within the world economy. In geographical terms, the world economy consists of a set of dominant core states in which most innovation and advanced economic activity takes place, and a dependent periphery and less dependent semi-peripheral area characterised by low-level production, low wages and coerced labour (Figure 1.1). Countries in the core have relatively high incomes, advanced technology and diversified production. They are generally prosperous and their populations enjoy high standards of living and qualities of life (Knox and Agnew, 1994). Those outside the core are less well developed and have economies that depend upon primitive technology and undiversified production. Many, especially in the periphery, are poor and are severely deficient in infrastructure and social provision. World systems theorists argue that the core achieves its greater prosperity through economic and political domination and control over peripheral and semi-peripheral areas. This relationship and the geographical consequences which follow from it are maintained and perpetuated over long periods of time through the evolution and changing organisation and spatial relationships of capitalism.

The world economy is organised around and through cities. Implicit in the global approach is the view that cities must increasingly be seen as interacting and interrelated elements within an urban hierarchy that underpins and makes possible processes of capitalist accumulation and reproduction. Rather than merely acting as points of exchange for goods and services produced and consumed in their surrounding

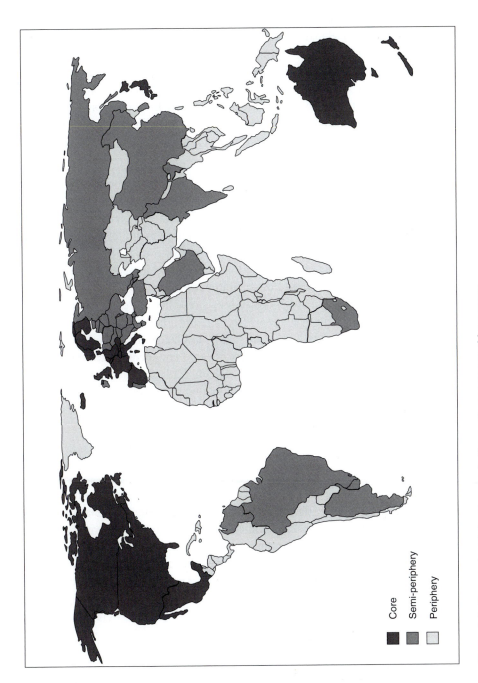

Figure 1.1 *Core and peripheral areas in the contemporary world economy*

Core

Semi-periphery

Periphery

areas, which was the historical pattern, they are places of articulation where people and products link to the wider world. Cities are locations through which global goods and services reach their markets and are consumed and from which surplus values are extracted. They interlock and intermesh in the form of local and national networks that in turn are incorporated within a global urban hierarchy. This system is dominated by a small number of world cities, housing the headquarters of the principal transnational corporations, finance and producer service organisations. Such centres are effectively the command and control points for global capitalism.

The emergence and spatial organisation of the urban world, according to this perspective, are dictated by the needs of the world economy. The accumulation of wealth through manufacturing, exchange and consumption is the primary cause of urban growth and urbanisation. It leads to a concentration of population in towns and cities throughout the core and in the periphery, so that urban development in both is an interdependent outcome of the operation of global capitalism. Spatial and temporal variations in levels of urban development are consequences of the ways in which capitalism has evolved and of its changing relations with areas of supply and demand across the world. Overseas resources and markets are secured and manipulated by colonisation, colonialism and imperialism. Explanations of urban development lie in the social and economic characteristics of successive forms of capitalism.

A global society?

A global society is emerging, it is argued, alongside and complementing the world economy. Its key features are ways of life and behaviours that are followed by increasing numbers of people around the world, irrespective of their backgrounds and places of residence. These are products of, and are regulated by, values, beliefs and aspirations that transcend geographical scales and spaces. They are reflected in common concerns for freedoms, tastes, fads and fashions. The world, it is argued, has converged in social terms at the expense of local differences. Globalism is replacing parochialism.

The emergence of global society is made possible by advances in transport and communications that overcome barriers of inaccessibility and distance and facilitate easy and cheap worldwide movements of ideas and people. Social networks that once were closed and localised have become open and interconnected, fusing at broader spatial scales. Global society is reflected in the long-distance and instantaneous circulation of information and imagery by broadcasting, telecommunications, videos and the World Wide Web of interconnected computers. It is reinforced by the international traffic in tourists, business travellers, migrants and workers. Such exchanges and flows bind individuals into interest groups and communities that extend across national and regional boundaries, and ethnic and cultural differences, and may be global in extent. They facilitate and encourage the creation of identities and relationships that are not proscribed by geography. Global society consists of spaces of flows, linkages and connections rather than spaces of places, such as states, regions and neighbourhoods, as was the case half a century ago (Castells, 1996). It exists independently of space. Membership and role are functions of participation and not place.

A strong case can be made that global society is increasingly urban in character. Cities are points of production and reproduction of urban ways of life and culture. As major and dense concentrations of population drawn from many different backgrounds, they are places in which a diverse array of beliefs, styles, values and attitudes originate, ferment and flourish. These combine, it is argued, in the form of patterns of association and lifestyle that are distinctively urban in character and differ markedly from those which exist in rural areas. Such modes of thought, relationship and behaviour are carried and spread by movements of people and flows of information and ideas well beyond city boundaries, so that they influence and may be adopted by populations across the world. Society is becoming urbanised in the sense that increasing numbers of people are being exposed to, and are absorbing, the social values that arise out of, and are most closely associated with, life in cities. Some of the key debates in urban geography concern the nature of global urbanism and the ways and extent to which it interacts with, and so may be destroying, modifying or reinforcing, traditional local cultures.

Globalisation debates

The globalisation approach must not be accepted uncritically despite its current popularity in the social sciences. Globalism is a contested concept, the existence and meaning of global economy and society being widely discussed and questioned by analysts (Johnston *et al.*, 2002). Opinions vary because of differences of emphasis and perspective and the difficulties of specifying and measuring many of the processes and patterns that are involved. Dissension and debate surround any intellectual conception and are the means through which ideas are tested and refined and explanation and understanding are advanced. They are especially wide-ranging with globalisation because it is such a big idea, a 'grand theory' or meta-narrative, which seeks to account for many, indeed most, of the social, economic, geographical and political characteristics of the contemporary world.

A central issue surrounds the extent of economic and social fusion. Globalists argue that most places are locked into worldwide economic and social networks which transcend national and regional boundaries. They point to the unlimited reach of air travel, the mass media and the Internet as evidence of interconnection and interdependency, and maintain that the consequence is growing economic and social coalescence. Few areas, it is suggested, are unaffected by these exchanges, so the adoption of a global perspective is justified. The counter-argument is that many countries and parts of countries have poor transport and communications and so can engage readily to only a limited extent with places outside the local area. Large tracts of Africa and Asia are without metalled roads, let alone Internet, mobile phone and television services. They are characterised by traditional subsistence or semi-subsistence rural economies and societies that function independently of national and global relationships. They exist in semi-isolation, being largely untouched by outside events. Major cities may be well connected, but these are few and far between in areas in which the village is the dominant form of settlement. A low level of global engagement is, similarly, a feature of many parts of communist and former communist countries, though for different reasons. The world, according to this perspective, may be coming together in economic and social terms for some, but for most it remains far apart; indeed, in relative terms the gap is widening. The concept of globalism is, therefore, at worst erroneous or at best premature.

A second area of contention surrounds the extent and meaning of global economic influences, even in those areas that are well connected. It is closely related to transnational corporations and their role and importance in the world economy. Globalists cite the proliferation of such companies and the spread of their products and brands as evidence of global economic power and domination. Such is their influence that transnational corporations are the focus of much criticism by opponents of global capitalism, who blame them for creating and perpetuating poverty and environmental degradation across the developing world. Critics of globalism, however, point out that transnational corporations make most of their money from the rich nations of Europe and North America and from customers who are well able to resist exploitation. An example is Nike that markets its products worldwide, but has only 20 per cent of its sales in the developing world. What seems to be a global corporation is, in fact, it is argued, a localised operator. Transnational corporations may have a global reach, but most are located in and serve developed world markets.

A third debate is grounded in history and concerns the novelty of global perspectives. A case can be made that global relationships are not new and were well established by the end of the nineteenth century, thus calling into question the present preoccupation with all things global. Proponents of this position point to the influence of the imperial powers at the time, and the free and widespread movement of capital, profits and labour around the world. The 'world' was much smaller but was as well connected as is the wider world today. Globalists counter this criticism by highlighting the distinctiveness of contemporary worldwide economic and social relationships and institutions. They argue that global patterns today are of a different kind and on a larger scale than any which existed in the past. Global exchanges involve flows of ideas as well as raw materials and manufactured goods and are mediated by global organisations and institutions on a worldwide basis to an extent that has never occurred before. The importance of historical antecedents is accepted, but the nature of contemporary globalism defines a new field of enquiry. It necessitates the adoption of new perspectives and the search for new understandings and explanations.

These contrasting viewpoints reflect extreme positions that are separated by degree as much as by substance. Few analysts reject the notion of globalism outright, although its significance and value as an intellectual construct is widely challenged. Debate is hindered by the paucity of global data, which means that theoretical arguments can rarely be tested empirically. In focusing upon urban outcomes, succeeding chapters evaluate the concept of globalism in detail and review the research and statistical evidence to support these different points of view.

Conclusion

This book examines the cities of the world and the world as a city.
It seeks to identify and account for the growth and organisation of the
contemporary urban world and the causes, characteristics and consequences of global urban society. Emphasis throughout is placed upon the
global patterns that appear when data are compared and mapped nation by
nation, and upon the worldwide processes which, many analysts argue,
are responsible for them. Variations in levels of urban development and
urban character within nation states, which of course may be pronounced
in large countries, are a focus of secondary attention. The approach is
largely synoptic, the purpose being to draw together, overview and
synthesise the literature, recent and established, so as to show how an
understanding of the urban world can be approached and has evolved.

Given the highly ambitious goals of analysing and explaining the
location and behaviour of half of the world's population, the treatment is
necessarily and deliberately broad. The aim is to provide an introduction
to the urban world and the global city by illustrating and exploring the
underlying processes, relationships and issues that constitute the subject
of more detailed and advanced study. Many questions are raised, but
few definitive answers are provided, this reflecting the innate difficulties
of generating explanations and understanding at the world scale and the
variable quality and availability of published research on global urban
issues. Individual chapters advance the overall aims by focusing in turn
upon historical patterns of urban growth and urbanisation (Chapter 3),
urban development as a global phenomenon (Chapter 4), some of the
socio-economic consequences of global urbanisation (Chapter 5), urban
culture and global urban society (Chapter 6), world cities (Chapter 7)
and the future urban world (Chapter 8). The definitions and data that are
so critical in urban analysis and upon which global urban study is based
are discussed in detail in the Appendix. Having outlined the aims and
objectives in general terms, Chapter 2 now provides an overall context
by using simple census statistics and elementary theory to outline and
explain the contemporary distribution of urban populations and urban
places at the global scale.

Recommended reading

Haggett, P. (2001) *Geography: A Global Synthesis*, Harlow: Prentice Hall.
A massive and masterly general overview, in 24 chapters, which integrates

human and physical geography in a global context. There are especially relevant sections on globalisation and urbanisation.

Held, D. (2000) *A Globalizing World? Culture, Economics, Politics*, London: Routledge. This volume examines the nature and meaning of globalisation, the ways in which it works and its consequences. It is especially strong on the globalisation of society and culture. As an Open University text, the material is presented in a form that is lively, interesting and easy to grasp.

Hubbard, P., Kitchen, R., Bartley, B. and Fuller, D. (2002) *Thinking Geographically: Space, Theory and Contemporary Human Geography*, Continuum: London. An excellent and detailed review of the history and contemporary status of geographic thought. The book includes profiles on key theorists and includes useful chapters on geographies of money and globalisation.

Johnston, R. J., Taylor, P. J. and Watts, M. J. (eds) (2002) *Geographies of Global Change*, Oxford: Blackwell. An invited collection of papers by leading geographers on a wide range of aspects of global distributions and global change in the late twentieth century. The editors provide useful contextual introductions and conclusions.

Knox, P. (1994) *Urbanization: An Introduction to Urban Geography*, Englewood Cliffs, NJ: Prentice Hall. A basic introduction to urban geography, using American examples. There are especially useful chapters on urbanisation and urban geography, and on the foundations of the American urban system.

Short, J. R. and Kim, Y. (1999) *Globalisation and the City*, Harlow: Pearson. An introductory text that examines the urban impact of, and the role of cities in, globalisation. The treatment is thematic rather than discursive, focusing upon debates about the meaning of 'global' and on economic, cultural and political aspects of globalisation and the city.

Topics for discussion

1 What do you understand by a 'geographical perspective' on economy and society?

2 What contribution can geographers make to the understanding of urban places and patterns?

3 What do you understand by globalisation? In what ways can the progression towards globalisation be measured?

4 'Globalisation is of little value as an explanatory framework in the social sciences'. How far do you agree?

2 Urban populations and places

By the end of this chapter you should:

- understand the economic and social reasons why cities exist;
- be familiar with the contemporary global distribution of urban populations and places;
- understand the debates which surround the nature of urban hierarchies and the reasons for their existence;
- be aware of the causes and consequences of primate city size distributions.

Introduction

The urban world is a heterogeneous place. Although in geographical terms the population of the globe is more urban than rural, levels of contemporary urban development vary widely. Important and highly significant differences exist within and between regions and countries in the size and proportions of their populations that live in urban places. The ways in which the population is distributed according to the number and size of cities also contrasts markedly from place to place. Such differences are the products of the many complex processes that are responsible for the contemporary pattern of global urban development. They have important implications for the ways in which the world integrates and functions as an urban system.

Generalisations concerning the urban world depend for their validity upon data that relate to each of the world's sovereign states. They are highly sensitive to definitions of urban places, the size and shape of countries and the timing and accuracy of national population censuses. Particular problems surround the reliability and frequency of censuses in many of the world's poorest nations where, paradoxically, the scale and rates of contemporary urban change are the greatest and the urban problems are the most severe. They are especially acute in the highly populated countries of the developing world, such as China and India, where even small errors of specification and enumeration may result in the misclassification or omission of many millions of people.

The quality of censuses throughout much of Africa is generally poor, so the numbers and rates of growth of the urban population are in many cases matters of informed speculation. International urban statistics are surrounded by many difficulties of availability and reliability and must be regarded as crude estimates rather than precise measures. Rather than refer to data problems continually throughout the text, they are examined in detail in the Appendix.

Some indication of the difficulties that can arise is shown by the statistics on urban China. The definition of 'urban' as followed in the State Statistical Bureau's *Statistical Yearbook* is based upon a range of criteria which includes place of residence, length of residence, place of population registration, the status of suburban counties and an individual's source of grain supply (Goldstein, 1989). A characteristic of Chinese cities, however, is that they frequently annex adjacent rural districts into their administrative areas in order to ensure control of vital urban supply needs, such as reservoirs or power plants. This wider definition gives a very different picture and accounts for the widely varying estimates of the size and percentage of the Chinese population that is urban.

Although it is important to be aware of the major shortcomings, there is little that can be done to compensate for deficiencies and to allow for differences. The most sensible course of action, and the one followed in this book, is to seek consistency by drawing upon a single data set with known characteristics. Detailed information on the urban world in the form of population estimates is assembled and published on a regular basis by the United Nations in its annual *Demographic Yearbook* (United Nations, 2003) and in its biennial *World Urbanization Prospects* (United Nations, 2002), which is available on-line. These incorporate data for each of the world's sovereign states that are principally based upon national censuses. It is by reference to these data, and especially the latter series, that the key features of the urban world, and the issues that surround them, are most easily and reliably identified and explored.

The contemporary urban world

The geography of the contemporary urban world is characterised by pronounced variations in the number and proportion of people who live in urban places. Some parts of the world have huge numbers of urban people; in others there are very few. The map of urban population, like the map of total population, is dominated by China and India (Figure 2.1). China has by far the largest number of urban dwellers and, as such, merits particular attention in any balanced analysis of the contemporary urban world. One in five of the world's urban people lives in China and the total population of Chinese towns and cities, at 460 million (in 2000), is similar to the urban population of Africa and South America combined. Chinese cities are both numerous and large, there being 28 with populations in excess of two million and 46 with between one and two million (Figure 2.2). The urban population of India is some 280 million, which is almost exactly the same as the urban population of the whole of Africa.

The urban populations of China and India are so high that they completely overshadow and suppress the more subtle variations which exist elsewhere and which may be highly significant at the regional scale. It is impossible to portray the urban populations of China and the Gambia on the same map! Sizeable urban populations occur in the USA, the former USSR, Mexico and Brazil because these countries are large and because a high proportion of their populations is urban. Japan and Indonesia are smaller countries with large urban populations. An urban population of some 549 million is spread across the 42 small nations which comprise the continent of Europe. Comparatively few people, however, live in towns and cities in Africa. This is both because of sparse population overall and because the percentage of the population that lives in urban places is low.

The distribution of the world's urban population is not reflected in the balance of research activity in urban geography. Most research has been undertaken on the towns and cities of the USA and Europe, although the majority of the urban population live outside these areas. It has generated a body of theory and concepts that are mostly based upon, and so inevitably to a degree reflect, western values and experience. The urban geography of China is especially under-researched and knowledge of Chinese cities and the ways in which they function and grow is fragmentary. The reasons are to do with the restrictions that are placed upon social and economic study in a totalitarian state and the paucity of published data that are available to researchers outside China. Until recently China was closed to foreign academics. It was not until the borders were opened in 1978 that urban fieldwork was possible.
A complete list of annual figures for urban population compiled on a comparable basis was not available until 1983 (Sit, 1985). There is in

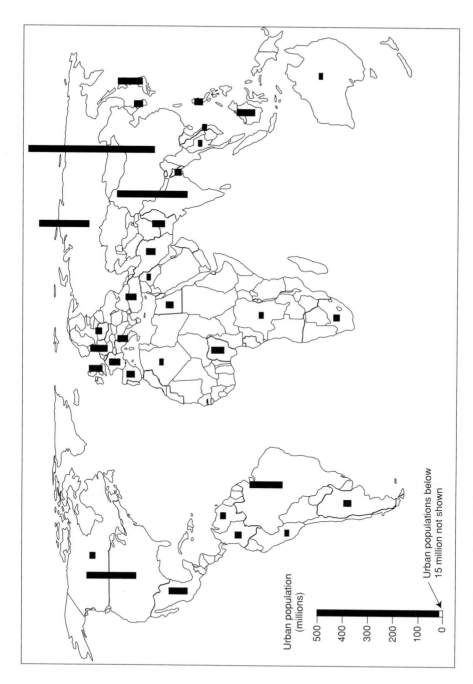

Urban population
(millions)

500
400
300
200
100
0

← Urban populations below
15 million not shown

Figure 2.1 Urban population, 2000

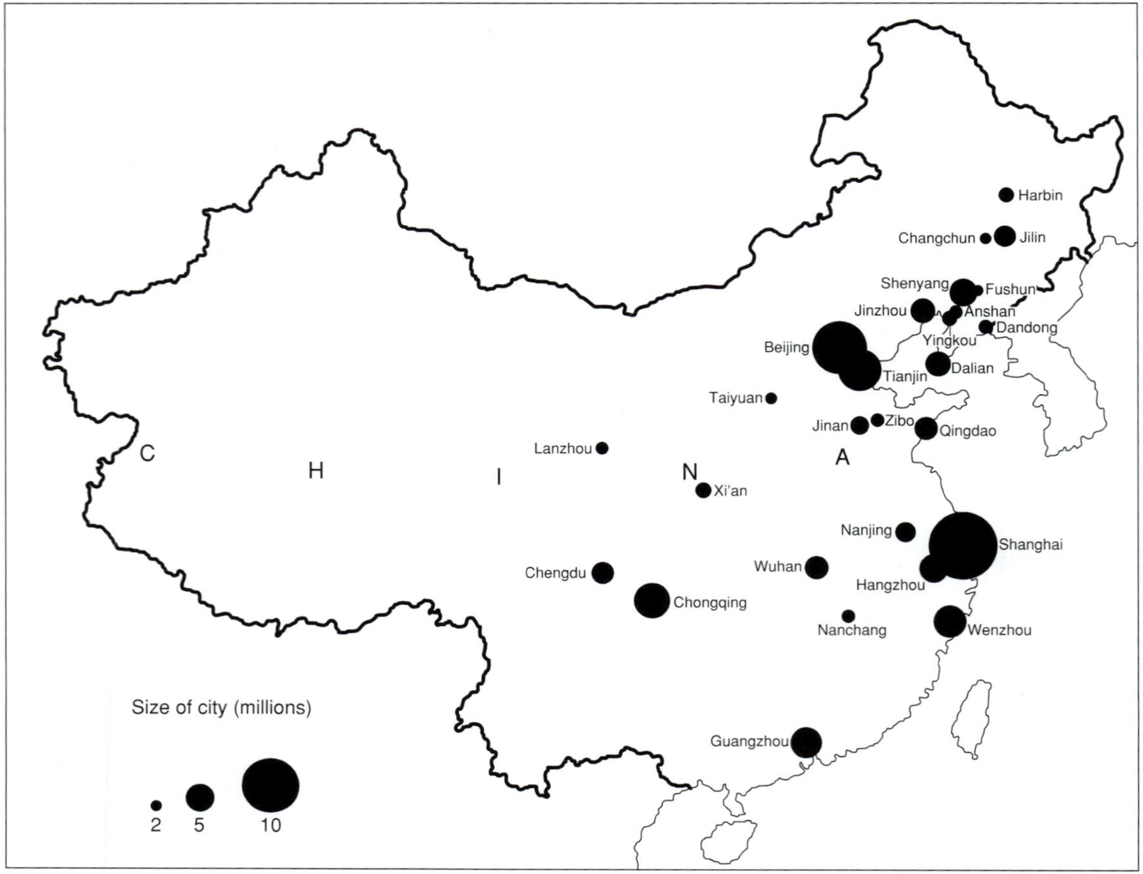
C H I N A
Harbin
Changchun
Jilin
Shenyang
Fushun
Jinzhou
Anshan
Dandong
Yingkou
Beijing
Tianjin
Dalian
Taiyuan
Jinan
Zibo
Qingdao
Lanzhou
Xi'an
Nanjing
Shanghai
Wuhan
Hangzhou
Chengdu
Chongqing
Nanchang
Wenzhou
Guangzhou
Size of city (millions)
2
5
10

Figure 2.2 ***Chinese cities with over two million people, 2000***

contrast a rich and long-established tradition of urban geographical study in India, based upon the excellent Census of India, which traces its origins to the work of R. L. Singh at the University of Delhi in the 1950s. The volume, however, is small in comparison with the size and scale of India's cities and urban problems. Such variations in levels of research effort detract significantly from the ability to form a balanced and informed view of the contemporary urban world.

Some of the highest levels of urban development in proportionate terms are found in South America, the most urban continent (Figure 2.3). The population is more urban than rural in all but one of the major South American countries (Guyana), and over 80 per cent of the population of Venezuela, Uruguay, Chile and Argentina are town and city dwellers. The proportion of the population that is urban is similarly high in Europe, Australasia and parts of western Asia. It is over 80 per cent in Belgium, Denmark, Germany, Luxembourg, the Netherlands, Sweden, the UK, Saudi Arabia, Israel, Australia and New Zealand. Belgium, with some 97 per cent of its population living in towns and cities is, with the exception of city states such as Hong Kong and Singapore, the world's most urban country. Unlike the Americas, where levels of urban development are uniformly high, there are, however, countries in Europe and western Asia that are predominantly rural. Examples are Albania, Moldova and Yemen.

Levels of urban development are low throughout most of southern Africa, eastern, south-central and south-eastern Asia. Only a small percentage of their populations live in urban places and these regions include many of the world's most rural areas. The village is the most common unit of settlement and towns and cities are the exception rather than the norm. Fewer than one person in three in southern Africa is an urban dweller. It is fewer than one in ten in Burundi and Rwanda and fewer than one in five in Burkina Faso, Chad, Eritrea, Ethiopia, Guinea-Bissau, Malawi and Niger. Countries in this region have small populations and the percentage that lives in urban places is low. The prevailing pattern in these areas is one in which the population is distributed so widely, and in such small settlements, that an articulated urban system, capable of supporting those modern facilities and services that require large concentrations, has yet to emerge (Rondinelli, 1989: 294).

The situation in eastern, south-central and south-eastern Asia is different because it contains the world's most highly populated countries, although the distribution of population within these countries, and in the

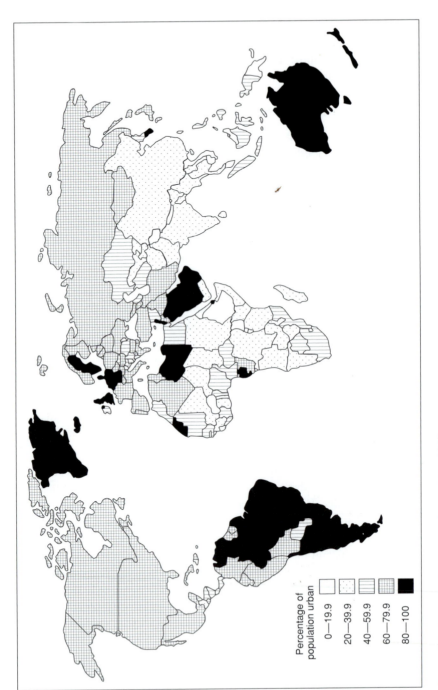

Figure 2.3 *Percentage of population urban, 2000*

Percentage of
population urban

0—19.9
20—39.9
40—59.9
60—79.9
80—100

region as a whole, is predominantly rural. With some 28 per cent of India's 1.1 billion people thought to be living in urban places, the level of urbanisation is broadly similar to that in China and Pakistan. Fewer than 25 per cent of the population of Afghanistan, Cambodia, Laos, Nepal, Sri Lanka and Thailand is urban. The island of East Timor in the Indonesian archipelago is reckoned to be the world's most rural sovereign state with only 8 per cent of its population living in towns and cities.

It is important to emphasise the complexity and variety of contemporary urban distributions and the fact that this undermines the value of generalisations concerning the urban geography of continents or regions. A particular complication is that they correspond only approximately with simple divisions of countries into 'developed' and 'developing'. The number and proportion of the population that live in towns and cities are products of a country's history, culture and resources and are only weakly linked to its level of contemporary economic development. The correlation between gross national product (GNP) per head of population and the percentage urban is 0.59 (Figure 2.4). Countries with high GNPs per head tend to have high levels of urban development, but there is a very wide range of levels of urban development among countries with low GNPs per capita. For example, the GNP per capita in Namibia and Peru in 2000 was below US$5,000, but the proportion of the population living in towns and cities was 78 per cent and 31 per cent respectively. Conversely, Austria and Ecuador have a similar proportion of their population in urban places (around 70 per cent), but their GNPs per head differ by US$18,000. The relationship is distorted by the fact that urban populations and levels of urban development are high throughout South America and yet the countries in that continent occupy only middle rankings on the World Bank's indices of economic development. South America has an urban history that is very different to that of other parts of the developing world. The data show that the developing world is highly differentiated in urban terms. There are as many variations of urban development within the developing world as there are differences of urban development between it and the developed world.

A world of towns and cities

The urban population is distributed among settlements of widely different size. Urban places range from towns and cities with several

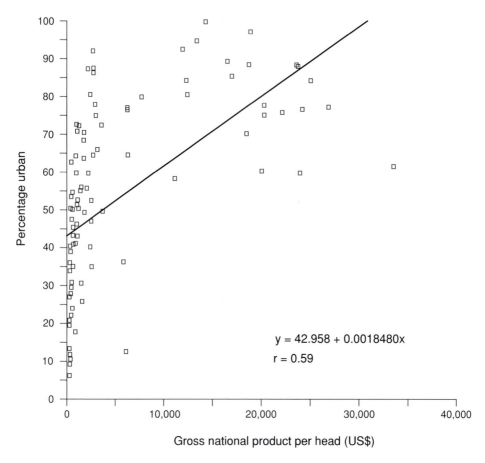

$$y = 42.958 + 0.0018480x$$
$$r = 0.59$$

Gross national product per head (US$)

Figure 2.4 *The relationship between urban and economic development*

thousands of people to those with tens of millions. An extensive
vocabulary of descriptive labels is available to describe these different
concentrations, although the terms lack precision and consistency of
meaning from country to country. It is difficult to make clear
distinctions in size terms, because urban places are members of a
continuum that grade into one another. Basic distinctions can be drawn
between towns and metropolises, and cities and megalopolises, but the
differences are not so sharp between adjacent members of the
continuum. Most would agree that settlements with populations over
100,000 are probably cities, but the status of places with around 20,000
is more questionable, especially when they have important local
government and commercial functions. As one goes down the scale from

the largest urban agglomeration to the smallest town it is extremely difficult to identify break points and terminology that are universally acceptable.

Most of the world's urban population live in small to medium-sized urban places. Despite the emphasis in the literature that is placed upon megalopolises and mega-cities, it is important to emphasise that the majority of the urban residents live in settlements whose populations are measured in thousands rather than in millions (Table 2.1). According to United Nations estimates, some 25 per cent of the world's population live in cities with fewer than 500,000 inhabitants. The figure is 75 per cent in the countries of the developing world. More detailed examination shows that the majority of the population in this settlement-size class, both in the world as a whole and in the developing world, live in cities with fewer than 100,000 people.

The primary function of most of these smaller urban settlements is to act as points of linkage between town and country, where agricultural surpluses are exchanged for urban goods and services. The importance of this role is reflected in the siting of such cities in highly accessible positions within local transport networks, and in their level and range of government, administrative and commercial functions. Such activities, and their spatial arrangements, are explained by central place theory, the general validity of which is verified by numerous studies of market centres and retail distribution networks and relationships across the world (Box 2.1). Some of the most important research questions in urban geography surround the ways in which these market centres are

Table 2.1 *Distribution of world population by size of settlement, 2002*

Size of settlement	Percentage of population
10 million or more	5
5 million to 10 million	3
1 million to 5 million	13
500,000 to 1 million	6
Fewer than 500,000	23
Rural areas	50
Total	100

Source: United Nations (2002).

Box 2.1 Central place theory

Central place theory seeks to explain the size, spacing and functions of service centres. It is grounded in an examination of the relationships between consumer demand and distance from a point of supply. Items such as food and drink are required daily and must be available locally, whereas consumers are prepared to travel greater distances for more expensive goods, such as cars and television sets, which they purchase infrequently. The theory integrates the concepts of threshold (minimum level of demand for profitable supply) and range (maximum distance over which purchasers will travel) for all the goods and services that are needed by consumers. The pattern for optimum servicing, combining most profitable provision and greatest access, is seen as one in which there is a uniform spacing and functional hierarchy of service centres. Towns and cities are equidistant from one another and are arranged in a series of tiers according to the variety and exclusivity of the goods and services that they offer.

Contemporary central place theory is the product of multiple enhancements to ideas originally proposed by Christaller (1933). It has been widely tested and has been shown to have a high level of validity, especially in areas with simple economies and easy movement (Beavon, 1977). Comparison of actual settlement patterns with those predicted by the theory enables planners to intervene so as to improve levels of servicing and efficiency. Central place theory provides a basis for planning the distribution of towns and villages in new areas of settlement, such as the polders of the Netherlands and the Gaza strip.

integrated within national and global urban systems and the consequences of such involvement for the lifestyles of their residents.

A notable feature of the contemporary urban pattern, however, is the degree to which the urban population lives in giant cities. In 2000 the United Nations recognised 320 cities with over a million people, which together housed one fifth of the world's population (United Nations, 2001a). There were 20 cities with between five and eight million residents. Some 4 per cent of the world's population lived in 24 mega-cities, or urban agglomerations, of eight million or more people (Figure 2.5). The best estimate is that Tokyo had some 26.5 million inhabitants in 2000, while São Paulo and Mexico City had 18.3 million and New York had 16.8 million (Box 2.2). The way in which the population is distributed among cities of different size has important geographical implications. Not only is the urban population concentrated in a small number of countries, but, within many of these countries, there is a disproportionate concentration in a small number of cities.

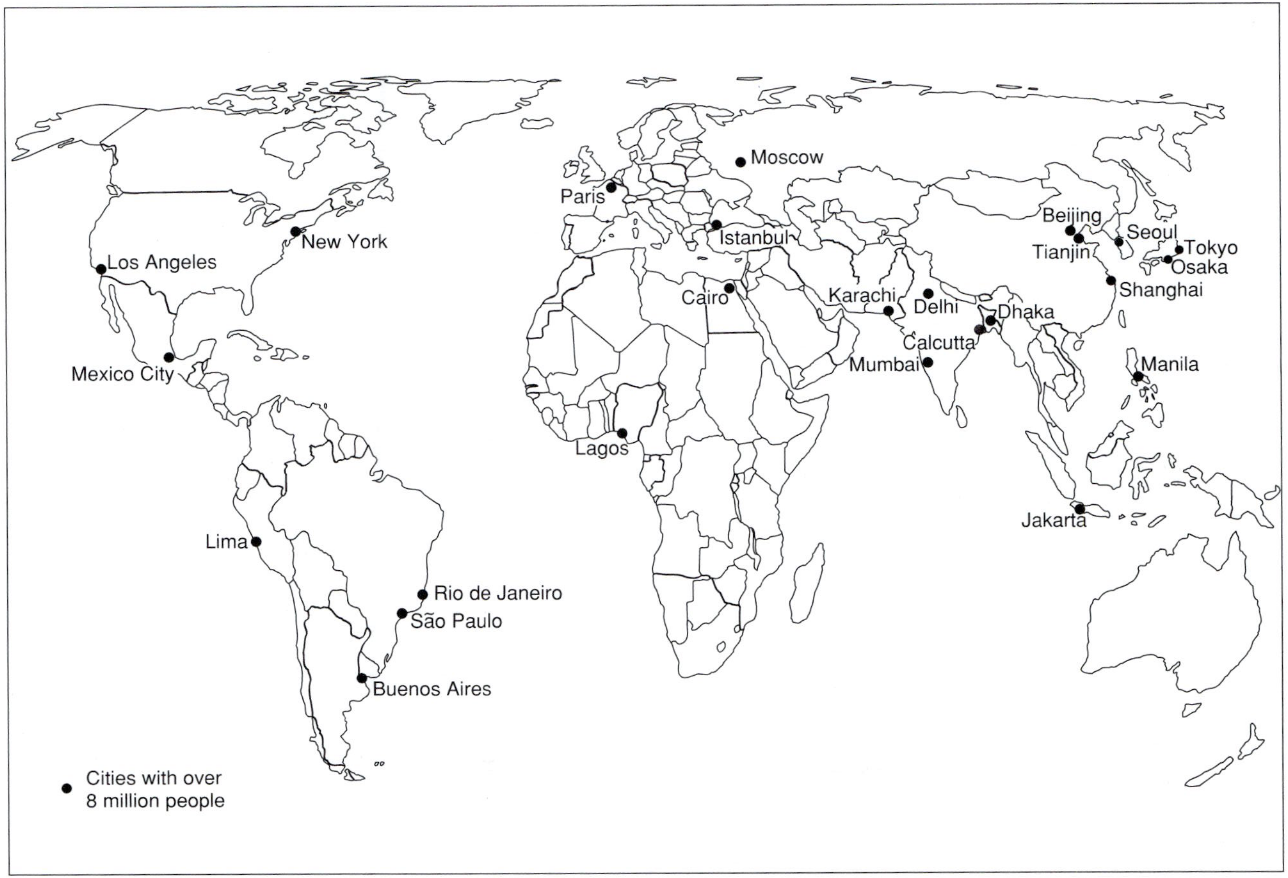

Figure 2.5 *Mega-cities, 2000*

Cities with over 8 million people

Los Angeles
New York
Mexico City
Lima
Rio de Janeiro
São Paulo
Buenos Aires

Paris
Moscow
Istanbul
Cairo
Lagos

Beijing
Tianjin
Seoul
Tokyo
Osaka
Shanghai
Manila
Jakarta
Dhaka
Delhi
Calcutta
Karachi
Mumbai

Box 2.2 Mexico City

Mexico City, with 18 million people, is a successful mega-city, although it has major social, economic and environmental problems. It is the principal economic and political centre in the country and has the greatest concentration of industry, the largest consumer market and the most highly trained workforce. Despite its size, it is a stable and relatively prosperous community. Some 80 per cent of homes have a piped water supply, and 70 per cent are on mains sewerage. Over 1,000 immigrants arrive in the city and are absorbed into the urban economy and society each day. Unemployment, underemployment, poor housing and congestion are significant problems, but they are less severe than in rural areas of Mexico. The city's principal difficulties, paradoxically, are physical rather than socio-economic, and stem from the poor initial choice of a site, at over 2,300 metres above sea level in an area prone to earthquakes (Griffin and Ford, 1993). It is also affected by frequent temperature inversions, which tend to cause a build-up of atmospheric pollution. Mexico City is far from being an urban utopia, but its existence and growth underline the viability and attractiveness of major concentrations of population.

The concentration of the urban population of countries into large cities occurs in all parts of the world. Again, it is a pattern that is independent of region, length of urban history and level of economic or urban development. Metropolitan dominance is most pronounced in South America and the Caribbean. The million cities in this region house 45 per cent of the urban population. Million cities equally dominate the urban hierarchy in many of the world's most rural regions and countries. Despite the low overall level of urban development, around 30 per cent of the urban population of south-central Asia lives in cities with over one million people. They include the mega-cities of Calcutta, Mumbai and Delhi. A similar percentage lives in million cities in southern Africa. Lagos is the largest city in this region, with a population of around 13 million.

Despite their enormous size, the world's major cities at present are viable and stable places that represent a significant social and economic achievement. They contribute disproportionately to national economic growth and social transformation by providing economies of scale and proximity that allow industry and commerce to flourish. They offer locations for services and facilities that require large population thresholds and large markets to operate efficiently. The major cities house many millions of people at extremely high densities and yet provide a range of opportunities and quality of life that is greater than that which is enjoyed in the surrounding rural area. Urban residents, even in the world's poorest countries, have better access to medical services (Plate 2.1) and higher levels of social welfare than those who live outside the city. The differences are reflected by most of the major social indicators. For example, infant mortality rates are lower in urban areas in 18 of the 22 developing countries for which comparable statistics are reported by Gilbert and Gugler (1992: 67). The urban rate is more than 20 per 1,000 lower than the rural rate in Brazil, Ecuador, Ghana, Indonesia, Liberia, Mali, Morocco, Nigeria, Peru, Senegal, Thailand, Togo and Zimbabwe.

Dire warnings of the imminent social or economic collapse of one or other mega-city appear periodically in the press, but such a disaster has yet to happen. Social conflict or economic catastrophes are normally played out across and between regions or nation states rather than exclusively within individual cities. Disturbing pictures of poverty, congestion and pollution in Calcutta, Mumbai, Rio and Bangkok, and of riots in Los Angeles and Beijing readily divert attention from what such places represent. Rather than gigantic social mistakes, cities, generally, are highly successful settlement forms.

Plate 2.1 *A doctor's surgery in Calcutta, India. In most countries there is better access to medical services in cities than in rural areas*

City size distributions and the global urban hierarchy

There are strong indications that global economic factors account for the way in which the population in a country is distributed among cities of different size. Implicit in the concept of the urban hierarchy is the assumption that population is spread across a range of cities of different sizes which interact and interdepend as a functioning urban system (Box 2.3). It was observed half a century ago by Zipf, among others, that, in many countries, there is a regular gradation of cities according to size (Zipf, 1949). In fact the second city was invariably, according to Zipf, half the size of the first, the third city a third the size of the first and so on, so that the size of any centre could be predicted by simply its rank and the size of the largest place. So widespread was this relationship thought to be that it became known as the 'rank-size rule'. The USA, the UK, Japan, China and Brazil are examples of countries in which the distribution of cities is approximately rank-sized (Brunn and Williams, 1993).

The situation in some countries is, however, very different in that the population is unevenly distributed among urban places. The extreme is reached when there is one excessively large primate centre that dominates all the others. Primate cities are not necessarily large in international terms, but they are, by definition, very much bigger than any other place in the country. They are most common and most pronounced among the poorest and most sparsely populated states in the developing world, although Portugal and Greece have primate patterns (Table 2.2). Primacy is especially prevalent in Central America and Africa. Such is the dominance of the primate city in some countries that it houses a disproportionate share of the national as well as the urban population.

Box 2.3 Poland

Poland is an example of a country in which there is a highly developed urban hierarchy and an even spread of cities. Warsaw, the largest city, with a population of 1.6 million, is about twice as large as the second city of Lodz (815,000) and there is a gradation of cities by size through Krakow (740,000), Wroclaw (641,000), Posnan (579,000), Gdansk (472,000) and Szczecin (419,000). The country is well integrated internally via extensive rail, road and air connections. Two-thirds of the population live in cities, but few areas lie outside the daily influence of an urban centre.

Table 2.2 *Levels of urban primacy, 2001*

Country	Largest city	Population in largest city as percentage of urban population
Guinea	Conkary	75
Panama	Panama City	73
Guatemala	Guatemala City	72
Lebanon	Beirut	70
Congo	Brazzaville	67
Dominican Republic	Santo Domingo	65
Kuwait	Kuwait City	62
Angola	Luanda	61
Haiti	Port-au-Prince	60
Portugal	Lisbon	60
Chad	Ndjamena	57
Cambodia	Phnom Penh	55
Thailand	Bangkok	55
Armenia	Yerevan	52
Afghanistan	Kabul	52
Burkina Faso	Ouagadougou	51
Costa Rica	San José	51
Greece	Athens	49
Côte d'Ivoire	Abidjan	48
El Salvador	San Salvador	48

Source: United Nations (2001a: Table B1).

Note: the table excludes city states.

Primacy has important spatial consequences. It means that, in many cases, most of the national population is concentrated in one small part of a country and the remainder is profoundly rural. Many primate cities are coastal ports, so that the population is peripherally rather than centrally located within the boundaries of the nation state. This applies to 14 of the countries in Table 2.2 and widens urban–rural differences still further. Such organisational and locational arrangements have far-reaching implications for internal integration and for the ease with which social and economic developments are likely to spread from the primate city to more distant regions (Box 2.4).

Research into city size distributions supports two broad generalisations. The first is that rank-size patterns are most common among mature, well-integrated and balanced economies, while primate distributions are a feature of embryonic urban systems and so are more common in the developing world (El Shakhs, 1972). The second is that the degree of primacy decreases over time: as the economy matures, so the population becomes more evenly spread across all the cities (Chase-Dunn, 1985). One city may remain dominant, but not excessively. Such findings suggest that city size distributions may relate in some way to the degree to which a country's cities are integrated within the global urban network.

A useful conceptual framework, first introduced by Berry (1961), is that of the urban system consisting of cities and the links between them. Such links may be flows of people, raw materials, goods, finance, information and ideas. At the global scale the urban system is closed, as it comprises every city and all their links. The urban system within each country, however, has varying degrees of openness because intercity links take place both within and across national boundaries. In some countries the urban system is wide open because its cities have many and varied links with places elsewhere. In others, for reasons of geography, history and politics, the urban system is effectively closed to

Box 2.4 Angola

The southern African state of Angola has a primate city size distribution, a poorly developed urban hierarchy and a thin spread of cities. The capital and largest city, Luanda, with a population of 2.7 million, houses 21 per cent of the country's 12.8 million and 61 per cent of the urban population. It has an international airport from which there are connections to other African capitals and to Europe. Location on the north-eastern coast, however, means that it has weak links with most of the country's 1.3 million sq.km of territory. There are only two railway lines in the country, both of which are presently closed, and few metalled roads. The next largest cities of Huambo (300,000), Benguela (185,000) and Lohito (160,000) are of local importance only. With only one-third of the population living in cities, most of the country is profoundly rural. The urban pattern in Angola is common across most of Africa and much of Asia.

most external influences and operates in relative isolation on a purely local basis.

The size distributional pattern is believed to reflect the number, direction and strength of the economic forces that act upon a country's cities. Primacy is held to be the simplest and most elementary arrangement, in which a few simple forces act strongly upon a single centre. Such a situation is most common in developing economies with relatively closed or isolated urban systems, where one place, commonly the national capital, is the focus of internal and international trade and so grows to dominance. At the other extreme, rank-size distributions are found where, because of the openness of the system, many forces affect the urban pattern in many ways. With greater economic complexity, there is a wider range of urban specialisms and more stimuli to urban development, all of which result in a rank-size pattern.

Smith (1985a, 1985b) suggests four reasons why primacy may exist in developing countries. The first is associated with colonialism and arises because empires tend to be controlled through key cities, which, as foci of imperial interchange, operate at a different and higher level than local or indigenous cities. Primacy is thus a function of colonial control, an explanation that appears adequately to explain the existence of dominant cities in Asia. This argument, however, is more difficult to sustain in South America, where colonial rule ceased much earlier and the urban hierarchy has had more time to adjust.

Primate cities are seen in the second interpretation as the major outlets for the products generated in dependent export economies. They are the points of linkage between interior producing regions and external overseas markets. As such, they benefit uniquely from processes of urban growth, which are associated with trade. An important consequence of modern means of transportation, especially railways, is that they funnel export commodities from producing areas to the primate city without contributing to the need for anything but very minor population centres in the hinterland, so creating a two-tier urban hierarchy and maintaining primacy. The importance of the railways in opening up interior areas of South America and in focusing development on the ports is well recognised, as Wilhelmy's work on Argentina shows (1986). The general applicability of export dependency explanations of urban primacy in the continent is confirmed by Smith (1985a), who found that, in seven of the eight South American countries studied, the level of export production correlated with the degree of primacy.

Smith's third argument is that primacy may be created from within by the collapse or decline of the rural economy. Local industry and trade are often destroyed by export dependence, and this undermines the economic base of provincial centres. In this case, the largest city grows at the expense of the smallest. Primacy, finally, may be a social consequence of the transition of an economy from subsistence to capitalist production. Such a change typically transforms class and labour relationships and, in particular, leads to a reduction in the amount of labour that is required in agriculture. Those who are no longer needed in farming tend to concentrate in, and so inflate the size of, major cities, where there are possibilities of jobs in service activities, or opportunities for income generation within the informal sector. Smith's research in Guatemala leads her to identify this process of labour release and urban concentration as the major reason for the continuation of primacy in parts of Latin America, where colonialism, export dependency and rural collapse no longer apply (1985b). Such are the incidence and persistence of primacy in the developing world that they point to the operation of several processes working in combination. Primacy may not have a single simple explanation, rather its origins are more likely to be multi-causal.

A basic problem with the analysis and explanation of city size distributions is that the size of many countries is poorly depicted by their population or geographical area. Indonesia, Japan and New Zealand consist of separate islands, which must to some extent function in isolation, whereas the elongated shape of Chile, Gambia and Laos inevitably affects the level of internal integration. Countries and political boundaries tend to change over time, so that it is difficult to relate urban size patterns to current national historical and economic circumstances. The evidence suggests, however, that primacy exists in countries in which the principal city is more strongly linked to, and integrated within, the global urban system than it is to the domestic urban hierarchy. This situation is especially common in colonies, former colonies and politically independent but economically dependent export economies.

Primacy points to the existence of a two-tier urban system in many of the countries of the world. People either live in one of the myriad small villages or in the primate city: there are few settlements of intermediate size. Primacy can be regarded as an indicator of 'over-urbanisation' in the sense that the largest city has far outgrown all the others and has become an atypical settlement in a particular country (Gugler, 1988).

Other commentators focus upon the lack of middle-sized settlements and the implications for the way in which the urban hierarchy functions. A deficient middle tier is a well-documented feature of the urban system in most African countries. For example, in 2000, only Algeria, Egypt, Morocco, Nigeria, South Africa, Zambia and Zaire had more than five cities, other than the largest urban centre, with a population greater than 100,000. Four of these countries accounted for nearly 70 per cent of Africa's secondary cities. Eleven other countries had only one or two secondary cities (United Nations, 2001a). Close parallels exist in India, although here the major shortfall is in the number of market towns that provide points of articulation between villages and the urban systems (Ramachandaran, 1993).

Differences in size are inevitably reflected in an imbalance of importance and role. Primate cities typically dominate their countries in economic and political terms. Invariably they are the centres in which national elites and other major decision makers and opinion leaders are concentrated. They are the headquarters for national television and telephone services and have the principal, indeed in many cases the only, international airport. In turn, they are the link points through which the dependent village population is connected to the global urban system.

Studies of city size distributions provide an indirect insight into the characteristics of urban linkages and connectivity. Together with work on the distribution of the urban population, they paint a picture of gross unevenness and variable integration. Not only are the spread of urban population and the level of urban development far from uniform, but cities differ significantly in the extent and ways in which they interconnect and interdepend. Cities in countries with rank-size patterns tend to be well integrated within wider networks; those in countries with primate distributions, with the exception of the primate city itself, are predominantly inward-looking and have strongest connections with the indigenous economy.

These differences suggest that the global urban system is presently fragmented and incomplete. Rather than a coherent whole, the contemporary urban world consists of a set of loosely knit subsystems. The largest is global in extent and is based upon movements and exchanges of people, goods, images, information and ideas among rank-sized urban centres and primate cities. It is dominated by the cities in the core economies, but also includes the primate and principal centres in the periphery. Other subsystems, in the periphery, have primate cities as their apexes, but have few external links except through the primate city and so function primarily at a local scale. Still others, in the most remote and backward parts of the periphery, are divorced even from national urban systems and so operate in comparative independence and isolation. Such observations provide support for those who question the idea of a global urban hierarchy and regard it as misleading and unhelpful.

Economic theories of urban formation

The world is an urban place because towns and cities offer substantial benefits over other forms of settlement. The advantages that people derive from clustering together are greater than when they scatter and disperse. Some of the most fundamental questions in urban geographical study are concerned with the precise nature and power of these agglomerative tendencies. Theories of urban formation seek to identify

Plate 2.2 *The economic and social bases of settlements. Towns and cities are supported by economic surpluses, some of which are generated in the surrounding rural area. They are products of powerful social processes which encourage people to concentrate for purposes of defence, security, religious practice and social interaction*

the forces that permit and encourage large numbers of people to concentrate in comparatively small areas. As such they form the bases of the theories of urban growth and urbanisation that are discussed in the next chapter. Two broadly contrasting viewpoints are prevalent in the literature, one underlining the primacy of economic benefits, the other emphasising the roles of the social bond.

Economic interpretations of urban formation lay stress on the savings of assembly, production and distribution costs that may be achieved through concentration. They argue that the existence of towns and cities is a consequence of the search for the most economical form of settlement. In primitive economies based upon labour intensive agriculture, the population is typically arranged in small village communities that are scattered across the countryside, a pattern that gives maximum access to the land. Towns and cities, according to this interpretation, come into existence when the level of agricultural production generates an annual food surplus (Childe, 1950). This critical development frees part of the workforce from agricultural employment and makes possible a range of craft and trade activities that cluster together in space so as to gain the maximum benefits of economies of scale and agglomeration. Cities are thus formed through the geographical concentration of surplus product (Harvey, 1973: 216). Activities that attract surpluses into the city are the core elements in the urban economy.

Underlying this economic interpretation of urban formation is a set of relationships that is best explained by economic base theory. At its most simple level, the urban economy may be viewed as two interdependent sectors, the basic and the non-basic (Figure 2.6). Cities can only exist at the expense of rural areas by buying in food and most raw material and product requirements, so any importing activity is 'city-forming' in that it makes possible and sustains the urban population. The basic sector consists of all those activities and employment that produce goods and services that are sold outside the city and provide the finance to enable basic requirements to be imported into the city. Corn and seed merchants, agricultural advisory services and farm machinery manufacturers which are urban-based and which serve a non-urban market are obvious examples of basic activities, but the classification also includes a wide range of manufacturers and service providers which 'export' their products across the city boundary. Although these city-forming functions are concerned exclusively with external markets, they themselves generate demands for goods and services for

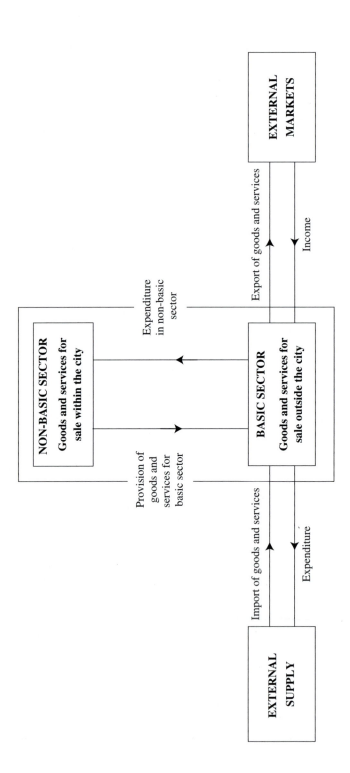

Figure 2.6 *Elementary structure of the urban economy*

their own support within the city. The non-basic sector consists of all those activities that provide goods and services for the city itself. Examples of 'city-serving' activities include municipal government; street-cleaning services; police, fire and ambulance services; corner stores; video hire shops and takeaway food outlets. Together the basic and non-basic sectors account for all the activities and employment in the city.

The two sectors are functionally interdependent. Any change in the size of one sector will be associated with a change in the size of the other. If the basic sector expands, workers in that sector will spend more on city services, so the non-basic sector will grow as well. Differences in the size of the two sectors, however, mean that changes in one will have a differential effect upon the other. For example, in a city with a basic:non-basic ratio of 1:3, an increase of 10 in basic employment will yield an increase of 40 (10 + 30) in total employment. The urban economic multiplier is a central concept in the explanation of urban formation and growth. It represents a mechanism whereby increases in the volume of external trade, and hence the size of the basic sector, result in a corresponding growth in employment in the non-basic sector, and thus an overall increase in employment and population in the city as a whole. In practice, the mechanisms of growth are of course highly complex. For example, if business output expands in response to increased sales, so profits and wages will increase, expenditure by workers and shareholders will rise, the demand for labour will grow and the population will reach new levels or thresholds resulting in the entry of new firms into the market (Figure 2.7). Urban growth is best seen as a circular and cumulative process, as growth in one sector triggers expansion through secondary multiplier effects elsewhere in the economy (Pred, 1977).

Studies of the economic structure of cities are difficult because many people's jobs involve both 'exporting' and 'city-serving' activities, so the analytical procedures that have been developed produce only crude breakdowns. The most widely used is the 'minimum requirements approach', as pioneered by Ullman and Dacey (1962). This involves classifying cities into size groups and then examining the percentage of the total labour force that is employed in each major category of occupation. The lowest percentage recorded for any city in each group is assumed to be the minimum necessary to enable cities of that size to function. These minimum requirements are equated with the non-basic workers and the number of workers over this minimum figure is taken to

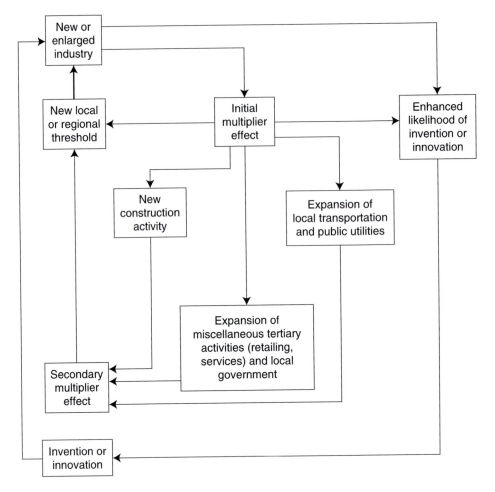

Figure 2.7 *The circular and cumulative model of urban growth*
Source: adapted from Pred, 1977.

represent the basic labour force. Empirical studies using this and related methods have identified the relative size of the two sectors. They show that the basic sector decreases in size as the urban population increases (Figure 2.8). For example, in a city of 10,000 people, approximately two-thirds of all employment is in basic activities, whereas in a city of 15 million, the figure is nearer one-quarter. A related finding is that the size of the multiplier similarly increases with urban size. It has a value of around 0.75 for a city of 200,000, but is nearer 2.00 for a city of six million.

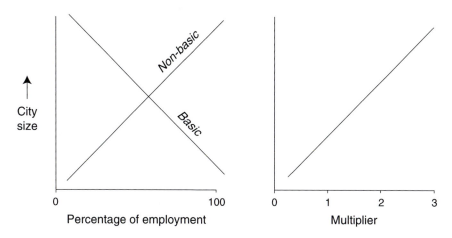

Figure 2.8 *General relationships between basic and non-basic components, and the urban economic multiplier, with city size*

Several important theoretical and practical implications follow from these findings. The first is that the larger the city, the less it is dependent upon basic activities, and hence its links with surrounding suppliers and markets, for its viability. The volume of external trade is critical for small towns and determines whether they expand or decline, but it is of secondary importance for the metropolis. Beyond a certain size, thought to be in the region of 250,000, growth is self-generated and is a product of the non-basic sector. The second consideration, which relates to the size of the urban multiplier, is that larger urban centres have the capacity to grow more rapidly: a small increase in the basic sector leads to proportionately large increases in the non-basic sector. Conversely, this means that large cities are somewhat more vulnerable to collapse if the basic sector suddenly contracts. These mechanisms and relationships emphasise the close dependence of small cities on the volume of surplus product that exists locally. They provide theoretical support for the views of Childe (1950) that the generation of surpluses was a key factor in urban genesis. Studies of urban economic bases raise questions concerning the optimum size of cities. There are important implications for planning, especially in areas where new towns and cities are being built from scratch.

It is important to emphasise that cities exist because of, and at the expense of, their surrounding environments. This applies to imports of food surpluses as summarised by elementary economic base theory, and, when their total environmental needs are considered, to their dependence on external sources of air, water and energy as well. The underlying relationship is fundamentally parasitic, since cities make no food, clean no air and clean very little water to a point where it can be reused.

An important practical consideration is the long-term viability of these relationships and its consequence for a sustainable urban future, issues that are considered in more detail in Chapter 8.

Social theories of urban formation

Although urban places need to import agricultural surpluses, it is, however, by no means certain that this was the formative event in the first emergence of cities.

> It has not yet been demonstrated clearly and unequivocally first, that a generalised desire for exchange is capable of concentrating political and social power to the extent attested by the archaeological record, or second, that it can bring about the institutionalisation of such power.
>
> (Wheatley, 1971: 282)

The fact that urban centres are present in a wide range of economies and cultures throughout the world suggests that origins of urban living are a product of human relationships and lie in the interpersonal ties that encourage people to congregate in space. Social explanations of urban formation stress the gregarious nature of human behaviour. They point to the complementary properties of links such as male and female, mother and child, sender and receiver, speaker and listener, and giver and taker, and argue that such bonds introduce strong centripetal tendencies among human populations. Even small groupings offer security, defence, self-help and the prospects of finding a mate, so they are attractive to non-group members. With increased membership, the value of community benefits grows. Cities emerge when social institutions and mechanisms, in the form of defence, administration, government and religion, are developed, which enable the population to live together in sizeable concentrations in space. Social organisation is, therefore, accorded priority over economic developments as the independent variable in urban formation.

These arguments are most closely associated with the work of Adams (1966) on the emergence of cities in early Mesopotamia. He maintained that the rise of cities was pre-eminently a social process, an expression more of changes in people's interactions with each other than with their environment. The novelty of the city consisted of a whole series of new institutions and a vastly greater size and complexity of social unit, rather than basic innovations in subsistence. For Lampard (1965) society has evolved through a number of organisational stages, each of which is associated with different settlement forms. Particular emphasis is placed upon the 'primordial' since this represents the achievement of a level of social organisation that is necessary to support and sustain village life. Improvements in agricultural productivity are an essential prerequisite, but the development of community structures through which it is possible to manage farming and ensure social stability is crucial to the viability of the settlement. The existence of cities, the next stage, represents the achievement of a higher level of social sophistication and consensus. This is reflected in formal bureaucratic, religious, military and political systems, which together enable large concentrations of people to coexist peacefully, harmoniously and prosperously. Underlying the argument is a strong emphasis on the urban consequences of social development. While the role of economic productivity in the rise of cities is not denied, its importance is held to be contingent upon the achievement of a given stage in social development. In this way, the city is 'a mode of social organisation which furthers efficiency in economic activity' (Lampard, 1955: 92).

Although social theories of urban formation were developed to account for the rise of early cities, they are of substantial relevance today. The development of appropriate social structures, organisations and institutions is as necessary for ensuring the stability and prosperity of contemporary mega-cities as it was to the viability of early agricultural settlements. The nature of urban exchanges, economic bases and multiplier effects is similarly of central importance in understanding the course of contemporary urban development. Cities exist on the basis of surplus product, whether food supplies that were of critical significance in antiquity, or manufactured goods and services that are the principal items of modern economic exchange. The ways in which such flows focus upon individual centres are primary determinants of the shape of national urban hierarchies and the arrangement of centres into a world system of cities.

Conclusion

This chapter has been concerned with basic distributions and elementary explanation: how many urban people there are, where they live, why they live there and the ways in which cities mesh together in the form of functional urban hierarchies. Attempts to overview patterns at the world scale, using nation states as the basic units of measurement, are difficult because of the underlying complexity, so there is inevitably a very fine line between useful generalisation and naive oversimplification. They are not helped by the wide variations that exist in the size and shape of countries and in the availability and quality of their census data. What the statistics show is that the world is now more urban than rural and that urban development is well established in most countries and regions. Despite the interest that is generated by million and mega-cities, most people live in small or intermediate centres whose size and role probably relate more closely to the productivity of their local areas than to the operation of global economy and society.

The interconnectedness and integration of settlements at the global scale is variable. Most major cities across the world have strong international links and interdependencies, but studies of city size distributions, and especially of primacy, point to many instances of weak connections with the lower order centres in the domestic urban hierarchy. The concept of the global urban system has considerable attraction and appeal as an intellectual construct, but its substance is contested. Many of the smaller places in the periphery, and the extensive areas which they serve, function in comparative independence and isolation.

The urban world is the product of processes of population concentration that began many millennia ago, when the generation of annual food surpluses and the creation of viable social structures led to and enabled the establishment of towns and cities. Although the underlying social and economic prerequisites are the same, the patterns outlined and explained in this chapter owe most to recent and continuing trends that operate at a global scale. It is only in the last half century that urban development has become sufficiently widespread across the world as to suggest causes that relate to the emergence of the world economy and society. Having taken a snapshot view of the reasons why cities exist, and of the contemporary urban system, it is now appropriate to explore the dynamics of global urban change by analysing processes and patterns of urban growth and urbanisation in detail.

Recommended reading

Brunn, S. D. and Williams, J. F. (1993) *Cities of the World: World Regional Development*, New York: HarperCollins. A set of specially commissioned chapters that examines levels of contemporary urban development on a region-by-region basis. Particular attention is paid to the principal cities in each area.

Carter, H. (1995) *The Study of Urban Geography* (4th edn), London: Arnold. A standard text in urban geography which includes a useful discussion of economic and social interpretations and explanations of the origins of towns and cities.

Dogan, M. and Kasarda. J. (1989) *The Metropolis Era, Vol. 1: A World of Giant Cities*; (1990) *The Metropolis Era, Vol. 2: Mega-Cities*, London: Sage. Volume 1 examines the characteristics and processes of urban growth in both the developed and the developing worlds. Volume 2 presents an in-depth analysis of ten giant cities around the world.

Gilbert, A. (1996) *Mega City Growth in Latin America*, Tokyo: United Nations University Press. A detailed account of the growth, characteristics and problems of the major cities of Latin America. Individual chapters focus upon Buenos Aires, Lima, Bogotá, Rio de Janeiro, São Paulo and Mexico City.

United Nations (2001) *Cities in a Globalising World: Global Report on Human Settlements, 2001*, London: Earthscan. An authoritative and up-to-date assessment of the condition of the world's cities and the prospects for making them better. The report comprises a set of interpretative papers and detailed statistics on the population, social and environmental characteristics of urban places.

Key web site

www.un.org This is the main site for the United Nations. It provides links, via 'economic and social development', to 'population', which gives access to the biennial reports on *World Urbanization Prospects*, and to 'human settlements', which gives access to the *Global Report on Human Settlements* and the *State of the World Cities* report.

Topics for discussion

1 What do you understand by the term 'global urban system'? Critically evaluate its validity by reference to the spatial and size distribution of towns and cities in contemporary Africa.

2 Outline and account for the characteristics of the contemporary global urban hierarchy.

3 'The generation of agricultural surpluses was the key development in the genesis of towns and cities'. Discuss.

4 Why is the urban hierarchy in many countries dominated by a disproportionately large city?

3 Urban growth and urbanisation: historical patterns

By the end of this chapter you should:

- be able to distinguish between urban growth, urbanisation and the spread of urbanism;
- understand the main ways in which urban growth and urbanisation are occurring;
- be familiar with the concept of counterurbanisation and understand its causes;
- understand the ways in which the distribution of population in and around cities changes over time;
- be aware of the major changes which have occurred in the course of global urban development and the reasons for them.

Introduction

The present pattern of global urban development is merely the most recent product of processes of urban change that began over 8,000 years ago. It represents an intermediate stage in the progression from a wholly rural to what will possibly be a completely urban world. The global urban pattern is changing in three different and unconnected ways through urban growth, urbanisation and the spread of urbanism. Urban growth occurs when the population of towns and cities rises. Urbanisation refers to the increase in the proportion of the population that lives in towns and cities. Urbanism is the name that is most commonly used to describe the social and behavioural characteristics of urban living that are being extended across society as a whole as people adopt urban values, identities and lifestyles. This chapter and Chapter 4 identify and attempt to identify and account for recent patterns of urban growth and urbanisation at the global scale. The origins and spread of urbanism are the focus of Chapter 6.

Urban growth and urbanisation are separate and independent trends. Urban growth refers to the absolute increase in the size of the urban population. It occurs both through natural increase, which is an excess of births over deaths, and through net in-migration. In most cities both factors operate together and reinforce each other, although the relative balance varies from place to place and at different times. Growth rates are compounded when the in-migrants are young adults. These are the most fertile age group and their influx is likely to raise the rate of natural increase. However, if the in-migrants are predominantly of one sex, the accompanying rate of natural increase is likely to be lower. As an actual rather than a percentage figure, urban growth is not subject to any ceiling. It can take place without urbanisation occurring so long as rural growth occurs at the same rate. It is likely to continue after urbanisation has ceased, as the population, which will all be living in urban places, goes on growing through the excess of births over deaths.

Urbanisation measures the switch from a spread-out pattern of human settlement to one in which the population is concentrated in urban centres. It is concerned with the relative shift in the distribution of population from the countryside into the towns and cities. Urbanisation is a change that has a beginning and an end, the former being when the population is wholly rural, the latter occurring when everyone is recorded as living in an urban place. In a world in which half the population presently resides in urban places, this latter situation is clearly some way off and is unlikely ever to be achieved. A sizeable rural population engaged in agriculture, fishing and forestry will always be required. There is growing evidence from contemporary experiences in developed countries that urbanisation tends to rise to a peak of around 85 per cent urban, and then falls slightly as processes of counterurbanisation take over. Models of urban spatial evolution suggest that levels of urban development of around 80 per cent are probably the optimum for sustainability.

It is important to emphasise that, as the total population of a country consists of both urban and rural dwellers, an increase in the 'proportion urban' is a function of both. It occurs when the urban component increases in relative size, either through faster urban growth or more rapid rural population decline. The measurement of urbanisation is not without its difficulties, as it depends upon the division of a country into urban and rural. It is affected by changes in definition and the classification of centres that are made from time to time by national census authorities. Such issues are addressed in the Appendix.

Urban growth

By far the most important characteristic of contemporary urban change is the sheer scale of urban population growth. The urban population rose by 576 million between 1990 and 2000. Urban growth correlates strongly with overall population growth, so it is not surprising to find that the greatest gains occurred in highly populated countries where large numbers were added to the national population (Figure 3.1). The urban population of China alone rose by 154 million over the decade. Major increases also occurred in India (63 million), Indonesia (31 million), Brazil (28 million) and Nigeria (20 million). Little or no urban growth took place in Europe, where national population levels are virtually static. For example, the urban population of the Netherlands rose by a mere 900,000 between 1990 and 2000.

Urban populations are growing rapidly throughout Africa and southern Asia (Figure 3.2). The rates are highest in parts of sub-Saharan Africa and western Asia. Rwanda leads the way with an annual average rate of growth of urban population, between 1995 and 2000, of 10.1 per cent, and high rates of growth in the same region also occurred in Kenya, Malawi, Mozambique, Uganda, Tanzania and Botswana. These countries are predominantly rural and have few sizeable cities, but they gained urban populations rapidly. Oman, Yemen, Nepal and Bhutan also had rates of urban growth in excess of 6 per cent over the same period. Growth in most countries is a product of a natural increase of the urban population compounded by net in-migration (Box 3.1).

Little or no growth is occurring in urban populations throughout most of the developed world. The average annual rate of urban growth in Europe was less than 0.11 per cent per annum between 1990 and 2000. In 15 European countries there was a slight fall in the urban population. Very low growth and decline are products of trends that are affecting cities of different size and do not mean that the urban hierarchy is unchanging. Generally, metropolitan centres are stagnant or are losing population, while towns and small cities are gaining. The most significant of these trends is the fall in the population of major cities (Box 3.2). Metropolitan decline in developed countries is very much a feature of the last 30 years. It was first recognised by Berry in 1976 and was confirmed by subsequent censuses for the USA and the countries of north-western Europe (Champion, 1989; Fielding, 1989; Cross, 1990). Metropolitan decline is well established in the UK. All of the largest 15 cities lost population during the 1970s and again during

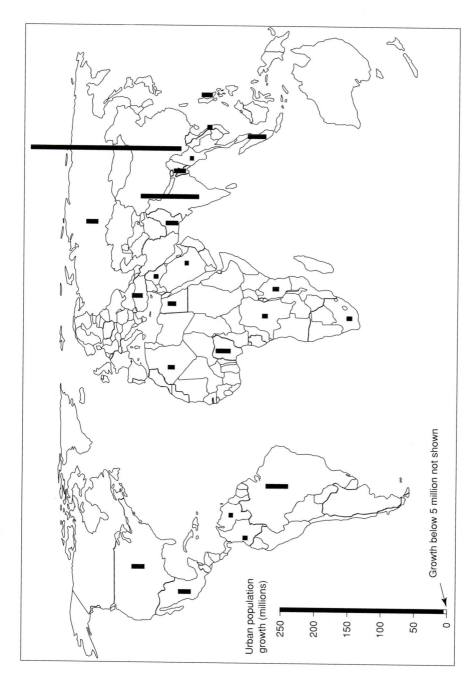

Urban population
growth (millions)

250

200

150

100

50

0

→ Growth below 5 million not shown

Figure 3.1 *Urban population growth, 1990–2000*

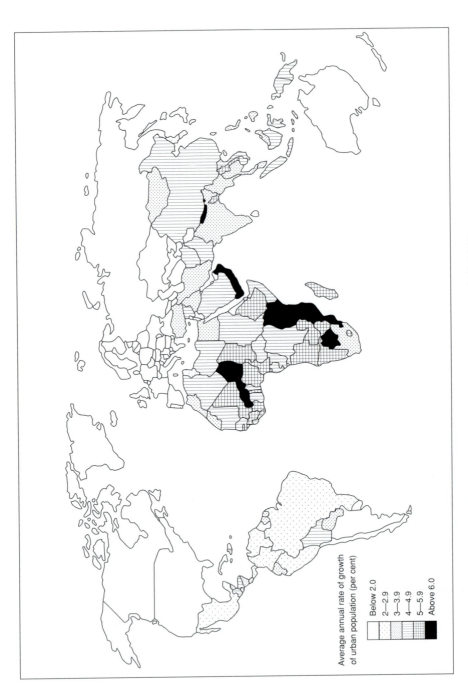

Average annual rate of growth
of urban population (per cent)

Below 2.0
2—2.9
3—3.9
4—4.9
5—5.9
Above 6.0

Figure 3.2 Average annual rate of growth of urban population (per cent), 1995–2000

Box 3.1 Components of urban population change

The population of a city is determined by the numbers of births, deaths and migrants. Natural increase (an excess of births over deaths) and net in-migration (more moving in than moving out) contribute to growth. Natural decline and net out-migration contribute to a fall in population. The balance of natural change and migration determines whether, overall, there is growth or decline.

Natural change and migration tend to be of roughly equal importance in determining the growth of cities in Africa and Asia. Natural increase was responsible for 61 per cent of the urban growth in the 29 countries studied by Preston (1988). In India it was 61 per cent during the period 1951–61, 65 per cent during 1961–71, and 53 per cent during 1971–81 (Bradnock, 1984). The figure was 66 per cent over the period 1975–90 in the 24 developing countries analysed by Findley (1993: 15)

Findley points out, however, that the migration component may be much larger because censuses fail to record temporary movers. A large number of migrants in the cities of the developing world circulate between urban and rural areas in repeated movements, some staying for a few months and others remaining for several years. Temporary or circular migrants are especially common in south-eastern Asia and western Africa, where they make up between 33 and 70 per cent of the total migrant population. On this basis, official statistics may miss one to two out of every three migrants.

Box 3.2 Components of urban population decline

Populations of many major cities in the developing world are stagnant or are in decline because of the combined effects of low natural increase and net out-migration. These reflect age structure and behavioural factors.

Births are falling because of a decline in the number of women of child-bearing age, and because they choose to have fewer children. The former relates to age structure and is a consequence of the relatively small number of females who were born 20 to 30 years ago (which itself is a function of the numbers born 20 to 30 years before that). The reasons for changes in reproductive behaviour are complex and include: changes in the roles of women; the availability of consumerist and careerist lifestyles as alternatives to traditional familism; and the consequences of marital breakdown. The increased availability of birth control enables this choice to be exercised. The number of deaths is rising slowly as the population ages.

More people are now leaving cities than are moving in. Age is a major factor as fewer people migrate to cities for work, and more retire (often early) and move out to more attractive areas. Increased car ownership enables more people to live in remote rural areas. With improved services and opportunities, rural areas are able to retain more of their population, especially young people who would otherwise have looked to cities for work.

the 1980s (Clark, 1989). Also, 13 of the largest 15 cities declined between 1991 and 2001 (Table 3.1).

An important corollary of contemporary urban growth at the global scale is the rapid increase in the number and size of the largest cities. Against the background of a general rise in the number of people who live in urban places it is the metropolitan centres that are proliferating and growing the fastest. United Nations estimates indicate that the number of cities with over eight million people increased from ten in 1970 to 24 in 2000 (Table 3.2). The number and size of mega-cities are increasing most rapidly in developing countries. In 1950, the only mega-cities, London and New York, were both in the developed world, while 18 of the 24 mega-cities in 2000 were in the developed world.

Table 3.1 *UK cities: population change, 1991–2001*

City	Population 2001 (000s)	Change 1991–2001 (000s)	Percentage
London	7,172	282	4.1
Birmingham	977	−29	−2.9
Leeds	715	−2	−0.3
Glasgow	578	−76	−11.6
Sheffield	513	−16	−3.0
Bradford	467	−8	−1.6
Edinburgh	448	27	6.3
Liverpool	440	−41	−8.6
Manchester	392	−46	−10.4
Bristol	381	−16	−4.1
Cardiff	305	5	1.8
Coventry	300	−5	−1.6
Sunderland	281	−16	−5.3
Nottingham	267	−14	−4.9
Newcastle	260	−19	−6.7

Source: 2001 Census.

Table 3.2 *Urban agglomerations with eight million or more persons, 1950–2000*

1950	1970	1990	2000
More developed regions			
New York	New York	Tokyo	Tokyo
London	London	New York	New York
	Tokyo	Los Angeles	Los Angeles
	Los Angeles	Moscow	Moscow
	Paris	Osaka	Osaka
		Paris	Paris
Less developed regions			
None	Shanghai	Mexico City	Mexico City
	Mexico City	São Paulo	São Paulo
	Buenos Aires	Shanghai	Shanghai
	Beijing	Calcutta	Calcutta
	São Paulo	Buenos Aires	Mumbai
		Mumbai	Beijing
		Seoul	Jakarta
		Beijing	Delhi
		Rio de Janeiro	Buenos Aires
		Tianjin	Lagos
		Jakarta	Tianjin
		Cairo	Seoul
		Delhi	Rio de Janeiro
		Manila	Dhaka
			Cairo
			Manila
			Karachi
			Istanbul

Source: United Nations (2001a: Table B1).

Urbanisation

The proportion, as well as the total number, of people who live in towns and cities is also increasing at the global scale. Urbanisation involves a significant shift in the distribution of population from rural to urban locations. Each year some 35 million more people are added to the world's towns and cities than to its rural areas. Although at 0.8 per cent per annum the global rate of urbanisation seems comparatively modest, it has profound implications for the long-term distribution of population. Around 25 per cent of the world's population lived in towns and cities in 1950. It is likely to be 65 per cent by 2020 (Figure 3.3).

Urbanisation is a cyclical process through which nations pass as they evolve from agrarian to industrial societies. For Davis (1969) the typical course of urbanisation in a given territory, in common with many growth processes, can be represented by an S-shaped (logistic) curve (Figure 3.3). The first bend in the curve is associated with very high rates of urbanisation, as a large shift takes place from the country to towns and cities in response to the creation of an urban economy. It is followed by a long period of consistent moderate urbanisation. As the proportion climbs above about 60 per cent, the curve begins to flatten out, reaching a ceiling of around 75 per cent. This is the level at which rural and urban populations appear to achieve a functional balance. Historical data suggest that the highest rate of global urbanisation occurred at the end of the last century, when the distribution of the population in what were then the leading industrial economies was switching rapidly from rural to urban. The rate at present is moderate and principally reflects the pace of urbanisation in developing countries.

At any one time, individual countries are at different stages in the cycle, so it is necessary, in making sense of present rates of urbanisation (Figure 3.3), to take account also of the overall level of urbanisation (Figure 2.3). Some countries, such as Mozambique, Burundi, Tanzania and China, are presently experiencing high rates of urbanisation as their populations switch rapidly from rural to urban locations. In others the rate is modest, either because the cycle of urbanisation is complete, as in the UK, Australia and Saudi Arabia, or because, as in India, Pakistan, Myanmar, Zaire and Somalia, it has only recently begun. The switch in the location of population from rural to urban is presently quickest in many of the countries of Africa and Asia. Several are going through the high-rate phase of the logistic cycle, where the social and economic

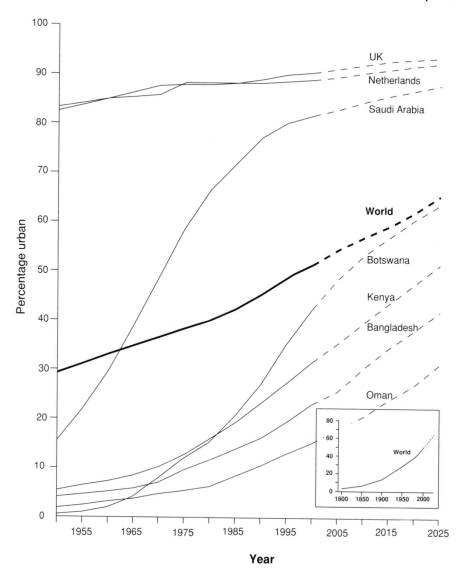

Figure 3.3 *Urbanisation trends, 1950–2025*

tensions associated with urbanisation are, in consequence, most severe. Further substantial shifts in the distribution of population can be expected in many countries with presently low levels of urbanisation as the proportion of the population that lives in towns and cities rises to ceiling levels.

Urbanisation is presently a developing world phenomenon. It involves the large-scale redistribution of people in many of the world's poorest nations that are least able to cope with its consequences. Urbanisation is occurring at greatest speed in countries that score low on the United Nations Human Development Index (Box 3.3) – their peoples have the lowest level of life expectancy at birth, and the lowest levels of education and per capita income (UNDP, 2002). It is however largely restricted to Africa and Asia. Little or no urbanisation is taking place in South America despite generally low levels of economic development, because the wholesale switch of population from rural to urban has already taken place. In examining the overall association between urbanisation and development, the World Bank estimates that urbanisation is increasing three times faster in low- and middle-income countries than it is in high-income countries. Because the rate is greater, levels of urban development in the developing world are catching up with, and will soon approximate, those in the developed world.

The geography of contemporary urbanisation is similar to, but is not the same as, the geography of urban growth. Countries that are presently urbanising most rapidly are mostly those in which there are the highest rates of urban growth (Figures 3.2 and 3.4). Rates of both urbanisation and of urban growth are presently highest in eastern Africa. The settlement patterns in this region are being transformed most radically through the combined effects of urban population increase and redistribution.

Little change is taking place in the urban and rural balance in the developed world because, in most countries, the cycle of urbanisation

Box 3.3 The Human Development Index

The Index is a measure of the 'quality of life'. It is calculated using three indicators: life expectancy at birth, educational attainment and standard of living. It was introduced by the United Nations in 1991 in an attempt to overcome some of the problems that were associated with purely economic measures of national development.

The Index uses rank order rather than absolute measures of level of development. In 1995, Canada had the highest HDI and was ranked number 1 in the world. At 174, Sierra Leone had the lowest ranking. The 15 lowest ranked countries were all in Africa, south of the Sahara.

**Average annual rate of
urbanisation (per cent)**

0–0.9
1–1.9
2–2.9
3–3.9

Figure 3.4 *Average annual rate of urbanisation (per cent), 1995–2000*

has run its course. More detailed analysis in fact suggests that, in many developed countries, the processes responsible for urbanisation have turned around. After many decades of expansion, major cities are in decline and population growth is taking place in rural areas. For example, nine of the 12 largest cities of Great Britain lost population between 1991 and 2001 and 11 of the 12 most rural counties gained, some at more than 5 per cent (Figure 3.5). This pattern was even more pronounced between 1981 and 1991 and was well established in the 1970s (Clark, 1989). The net effect is that there is a national deconcentration of population. For Berry (1976) this amounts to counterurbanisation (Box 3.4) in so far as the traditional processes that favour towns and cities at the expense of rural areas are now working in reverse.

Counterurbanisation replaced urbanisation as the dominant process of locational change in the USA more than three decades ago. Between 1960 and 1970 the metropolitan areas of the USA grew five times as fast as the rural areas. But during the 1970s the pattern was inverted, with the rural areas gaining population at one-and-a-half times the rate of that in the cities. A breakdown of the data revealed the extent to

Box 3.4 Counterurbanisation

Counterurbanisation is a shift in population at the macro scale from a state of more concentration to one of less concentration. It is a descriptive label that is applied to the observed pattern of change over time in the balance of urban and rural populations. No reasons for such changes are necessarily implied. The size of the population living in an area is determined by the numbers of births, deaths and migrants. Attempts to define counterurbanisation (or indeed urbanisation) in terms of any one of these components of change are in consequence misleading: counterurbanisation is not synonymous with net out-migration.

Counterurbanisation takes place at a broad (national) scale. There is a clear distinction between the long-standing and continuing process of central city loss and suburb/ hinterland gain, and the comparatively new phenomenon of counterurbanisation by which rural areas exhibit a resurgence of growth. The difference is that, whereas with a shift to the suburbs and exurbs the population remains within the daily urban system and so continues to participate in the routine life of the metropolis, with counterurbanisation there is a 'clean break' with the city. Counterurbanisation is in no way incompatible with decentralisation. Each refers to patterns of population change which may or may not be occurring at fundamentally different scales.

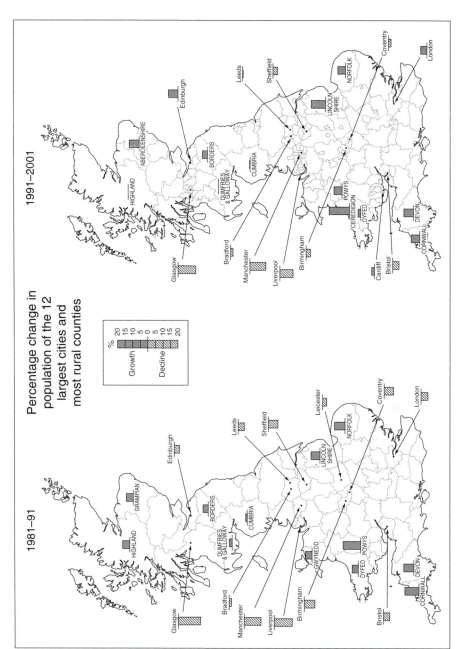

Figure 3.5 *Great Britain: percentage change in population of the 12 largest cities and most rural counties, 1981–2001*

which this change was attributable to counterurbanisation. The work of Berry (1976) showed that approximately half of the new non-metropolitan growth was adjacent to cities and amounted to no more than suburban sprawl across excessively tightly drawn boundaries. A roughly equal portion of non-metropolitan growth was, however, non-adjacent to, and remote from, existing cities and so represented true non-metropolitan revival. Counterurbanisation is seen by some as representing a distinctive and advanced stage of urban development that will eventually redress the imbalance in population between urban and rural areas. The pace of counterurbanisation has, however, recently slackened across much of the developed world, including Great Britain, so it could be no more than a minor blip in an otherwise continuing process of urbanisation (Champion, 1999, 2001).

Stages of urban development

If the western metropolis is the most advanced form of settlement, then counterurbanisation may represent a phase in the urban life cycle that will ultimately be followed by populations in what is presently the developing world. For Berg *et al.* (1982) cities evolve in a clearly defined sequence of stages that can be conceptualised in a model of urban development based upon population changes in urban regions as a whole and upon shifts of population within urban regions. A modified version of Berg's model is shown in Table 3.3. A fall in urban populations is most characteristic of the counterurbanisation stage that follows periods of urbanisation and exurbanisation. It may in turn be succeeded by an era of reurbanisation when cities return to a state of growth and expansion.

In the 'stages of development' model, changes of urban form are related to shifts in the distribution of population within and around the city. For this purpose it is useful to divide the urban landscape into a number of areas according to their population, employment and commuting characteristics. At the centre of the urban region is a core area of population, jobs and economic activity consisting of the physically built-up city. It comprises a central business district of shops and offices, surrounded by an inner area of mixed industrial, wholesaling, warehousing and residential uses. This is surrounded in turn by suburbs of estates and developments of successively newer and less densely packed houses.

Table 3.3 *Stages of development of a daily urban system (DUS)*

Stage of development	Classification type	Population change characteristics			
		Core	Ring	DUS	
I Urbanisation	1 Absolute centralisation	++	–	+	Total growth (Concentration)
	2 Relative centralisation	++	+	+++	
II Exurbanisation	3 Relative decentralisation	+	++	+++	
	4 Absolute decentralisation	–	++	+	
III Counterurbanisation	5 Absolute decentralisation	– –	+	–	Total decline (Deconcentration)
	6 Relative decentralisation	– –	–	– – –	
IV Reurbanisation	7 Relative centralisation	–	– –	– – –	
	8 Absolute centralisation	+	– –	–	

Source: modified from Berg *et al.* (1982: 36).

Note: the terms urbanisation, exurbanisation, counterurbanisation and reurbanisation are defined as follows: urbanisation occurs when the growth of the core dominates that of the ring, while DUS as whole is growing; exurbanisation occurs when the growth of ring dominates that of the core, while the DUS is still growing; counterurbanisation occurs when the growth of the ring dominates that of the core, while the DUS declines; reurbanisation occurs when the growth of the core dominates that of the ring, while the DUS declines.

Beyond the core is an extensive commuting ring from which it draws many of its daily workers. The ring encompasses an area of towns and villages in a predominantly rural setting in which the population focus their activities upon the core. The strength of the diurnal ties between the core and the ring means that the two areas together function as a daily urban system. They are tightly bound in a relationship of interdependency by morning and evening commuting flows and by movements between the two areas for shopping and recreation. Beyond the daily urban system is an extensive but sparsely populated rural area. As a relatively self-contained labour market, this region has no major urban centre of its own and so looks to the core on an infrequent basis and for only the highest-order services.

Berg's model of urban development is based upon variations in the direction and rate of population change between the core and the ring. Two types of change are recognised. Shifts are absolute when the directions of population change in the two areas are different, as, for example, when the core is growing while the ring is declining.

Alternatively, the shift is relative when each area has the same direction of change but the rate of change is different. Thus, a relative shift to the core would occur when both the core and the ring are growing but the population of the core is increasing at a faster rate. It is important to stress that the model is purely descriptive and makes no inferences as to how population shifts occur. Shifts of population arise because of differences between areas in the numbers of births, deaths and movements that take place. The relationships among these variables are typically so complex that simple associations, as for example between counterurbanisation and net out-migration, are likely to be misleading. Detailed analysis of the individual components of demographic change is necessary before reasons for observed population shifts can be advanced.

The urbanisation phase of urban development is the first of two stages that are characterised by the overall growth of the daily urban system (Table 3.3). It is associated with the expansion of employment opportunities in the city and with increases in the efficiency of agriculture that release workers from the land. The growth in the population of the daily urban system arises primarily because the core expands at the expense of the ring. Initially, the process of centralisation of population within the daily urban system is absolute, as the ring, which is still overwhelmingly rural, loses population to the core. It gives rise to a compact and densely populated physical city, as existed in Britain in the early nineteenth century and as is common in many parts of the developing world today. Later, as transport improvements allow some separation of place of residence from place of work, and as population spills over a tight city boundary, growth occurs in the ring. This results in a relative centralisation of the population because the rate of growth is less than that which is taking place in the core.

The most important feature of the 'exurbanisation' phase in the spatial evolution of the city is that the population decentralises. Although the daily urban system as a whole continues to increase in population, differences in the rate, and subsequently in the direction of change between the core and the ring mean that the daily urban system decentralises as it grows. The expansion of the ring reflects the environmental attractions of exurban locations and an increase in the number of people who can afford to move out of the core. It is facilitated by major transport improvements, including the development of suburban rail networks and the introduction of tram and bus services. Continued, though much slower, growth of the core means that a relative decentralisation characterises the initial phase of exurbanisation. Subsequently, as the core begins to lose population, the process of change becomes absolute. Core area decline and ring growth in the latter period are closely linked to rising levels of car ownership and use which enable large number of people to populate the commuter belt, especially those areas which are well away from the major radial routeways.

A decline in the population of the daily urban system, both core and ring together, distinguishes the third and fourth stages of urban development. In place of urban expansion, it is the rural areas beyond the daily commuting range of the core where growth takes place. The net effect is that there is a shift, at the national scale, from a state of more concentration or urbanisation, to one of less concentration or counterurbanisation.

In the stages of development model, absolute decentralisation characterises the first phase of counterurbanisation. With the daily urban system in overall decline, it is only the ring that continues to gain in population. Later, even this area declines and, as both core and ring lose population, the dominant form of change becomes that of relative decentralisation. Although counterurbanisation is now well established in many advanced economies, it represents only the most recent stage in the course of urban development. A continuation of the processes of deconcentration leading to more rural growth and urban decline is one possible future. Another is that of reurbanisation (Table 3.3). Berg *et al.* (1982) argue that urban renewal will eventually restore the appeal of the city. Although still declining overall, the daily urban system will, under these conditions, undergo a relative re-centralisation as the losses in the core diminish. With a return to limited growth in the core, the centralisation process becomes absolute.

The comparative recency with which powerful deconcentration trends have become established suggests that, for the foreseeable future at least, advanced western economies will be characterised by urban population decline and counterurbanisation. Urbanisation, however, as explained in the previous section, seems to be deeply entrenched throughout much of Africa and Asia, with few signs as yet of the onset of significant exurbanisation: significant daily commuting is not possible because of low levels of private car ownership and the limited capacity of feeder road networks. The stages of development model has no predictive content since it is purely a descriptive generalisation. Its principal value is that it provides a useful conceptual framework, based upon the experiences of western cities, with which to generalise and to pursue explanations about long-term changes in the distribution of urban populations at both national and intra-urban scales.

Reasons for urban growth and urbanisation

Urban growth and urbanisation at the global scale are very recent phenomena. Although towns and cities have existed since neolithic times, a massive rise in the number and size of cities and a wholesale shift of population from rural to urban have occurred only in the last 50 years. The world was very much a rural place in 1950, with only slightly over one-quarter of the population living in towns and cities (Figure 3.6). The population was more urban than rural in North America and parts of Europe, South America and Australasia, although only in the UK and the Netherlands did more than 80 per cent live in urban places. Most of Africa and Asia, and the remaining parts of South America, were rural with fewer than 20 per cent of the population being town and city dwellers. In these areas, such urban development as existed was highly localised and occurred, predominantly, in coastal pockets, so there were vast tracts of inland territory in which there was negligible urban population.

The 1950 pattern bears little resemblance to that of today. The differences between Figures 2.3 and 3.6 point to the operation, since the middle of the twentieth century, of new and powerful processes of population concentration which have extended urban patterns across the globe. As well as spatial extent, the speed of change is remarkable. It took nearly eight millennia for the population of the world to rise to 25 per cent urban: it took less than half a century to rise a further quarter to present levels. In order to understand the contemporary urban world it is therefore necessary to analyse the factors which led slowly and incrementally to the distinctive core/periphery pattern of urban development by 1950, and those responsible for global urbanisation since.

Urbanisation is the consequence of processes that concentrate people in urban areas at the expense of the countryside. Although towns and cities existed in many countries in the mid sixteenth century, overall levels of urban development were low. It is doubtful if any territory, or the world as a whole, was more than 1 per cent urban in 1550 and, of the few cities, only Paris, Naples, Venice and Lyons had populations in excess of 100,000 (Chase-Dunn, 1985: 279). Urban development was restricted

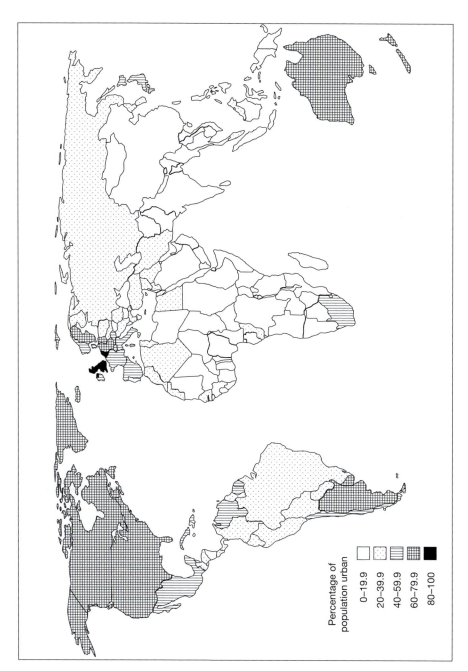

Figure 3.6 *Percentage of population urban, 1950*

Percentage of
population urban

0–19.9
20–39.9
40–59.9
60–79.9
80–100

by the volume and value of surplus product that could be generated and accumulated in a single place. The low level of economic development effectively imposed a ceiling on the number of people that could be sustained in urban places.

Most researchers agree that this constraint was broken irrevocably and irreversibly when industrialisation raised levels of output to unprecedented levels. This formative transition led to the production of huge surpluses and the consequent emergence of the first urban economy. Once self-perpetuating urbanisation had been 'invented' it was both copied and exported. The former process, which may be termed indigenous urbanisation, occurred in adjacent areas, where the social and economic circumstances were similar to those in the 'breakthrough' economy. It created a core region of advanced and dominant urban industrial economies. The latter process took place in those dependent territories that were most closely linked with the core. Significant time lags were involved in both processes, so gross unevenness characterised the early stages of global urban development.

The mechanisms involved can be explained by the interdependency theory of global urban development, so called because it sees urbanisation in both core and periphery areas of the world economy as interrelated consequences of a common set of causes (Clark, 1998). Global urban development, according to the theory, is a consequence of two linked processes: changes in the way in which wealth is accumulated, and the evolution of the world system of nations (Table 3.4). The former is a product of the sequential evolution of the prevailing economic formation from mercantilism, through industrial and monopoly capitalism, to transnational corporate capitalism (Castells, 1977; Slater, 2001). It has its own momentum in the form of the drive for ever higher levels of output and profit through the development of new sources of wealth and units of production. The latter structural development is concerned with the division of the world into progressively larger spheres of economic association and exchange based upon changing space relations and systems of supply (Taylor, 1993). It is associated with the rise to economic and political dominance of a small group of core nations, led by the USA as the foremost hegemonic power. These two sets of interdependent structural and spatial changes set in motion circular and cumulative processes of population concentration, in a manner described by Pred (1977) and outlined in Chapter 2, that affected the world at different times and in different places over the past three centuries.

Table 3.4 Principal stages in global urban development

	1500–1780	1780–1880	1880–1950	1950–
Mode of accumulation				
Economic formation	Mercantilism	Industrial capitalism	Monopoly capitalism	Corporate capitalism
Source of wealth	Trade in commodities	Manufacturing	Manufacturing	Manufacturing and services
Representative unit of production	Workshop	Factory	Multinational corporation	Transnational corporation, global factory
World system characteristics				
Space relations	Trade routes	Atlantic basin	International	Global
System of supply	Colonialism	Colonialism/imperialism	State imperialism	Corporate imperialism
Hegemonic powers	United Provinces, Mediterranean city states	Britain	Britain, USA	USA
Urban consequences				
Level of urbanisation at start of period (percentage)	2	3	5	27
Areas of urbanisation during period	European ports	Britain	North-western Europe, the Americas, coasts of empires	Africa and Asia
Dominant cities	Venice, Genoa, Amsterdam	London	London, New York	New York, London, Tokyo

Interdependency theory proposes a single explanation or interpretation for urbanisation, whether in developed or developing economies.
It has echoes in dependency theory, which explores and attempts to account for the links between development in core regions and underdevelopment in the periphery (Frank, 1967, 1969). Dependency theory suggests that underdevelopment is a result of the plunder and exploitation of peripheral economies by economic and political groups in core areas. Interdependency theory argues that urban development, wherever it occurs, is one of the spatial outcomes of capitalism. When seen from the developing world, most recent urbanisation appears to be 'dependent' in the sense that it is introduced or imposed by the developed world. From a global perspective, however, all urbanisation can be held to be interdependent in that it stems centrally from capitalism and its spatial relations. This is not to say that all urbanisation has arisen in an identical way and is therefore the same in all countries. Capitalism has adopted different forms at different times and is regulated in different ways, thus producing spatially differentiated patterns of urban development at the global scale.

The interdependency theory of global urban development can be criticised on four principal grounds. The first, in common with structuralist interpretations generally, is that it is stronger on suggested associations than on causal linkages. The fact that capitalism changed at a time of massive urbanisation does not necessarily imply a functional connection. Coincidence is not the same as causation and the mechanisms involved, which may vary over time and space, are matters for detailed empirical investigation and elaboration.

A second reservation is that urbanisation in the developing world lagged so far behind that in the developed world that it cannot be regarded as part of the same process. Britain was an urban industrial society for three-quarters of a century before any territory in what is now the developing world passed the 50 per cent urban threshold, and the urbanisation of most of the developing world did not gather real momentum until after 1950. It is important, however, to place urbanisation in its context of space and time. Global urbanisation involves massive shifts in the distribution of population over a wide area and is inherently a slow process. It is perhaps no accident that self-sustaining urban development first occurred in Great Britain, a very small country, where forces of urban growth and urbanisation were concentrated (Carter and Lewis, 1991). A sense of perspective is also important. When looking back over the last two centuries from the present, lags of a few decades appear to be of major significance. In the context of eight millennia of urban history they are trivial.

A third criticism is that interdependency theory undervalues the rich traditions of urban development, supported by non-capitalist economic systems that existed in many developing countries. Highly successful urban civilisations existed in ancient Egypt, India, China, Cambodia, Peru, Mexico and Nigeria in states and economic systems that were religious, military or feudalistic in formation. Independency theory however recognises the achievements of non-capitalist economies although it is argued that they were incidental to global urban development. Levels of productivity and surplus in early urban hearthlands were never high enough to facilitate self-sustaining urban development and so their importance was localised. Rather than denying and devaluing their contribution, interdependency theory provides a powerful explanation as to why non-industrial urban economies were not more successful.

The final criticism is that capitalist theories do little more than state the obvious and often in a language that serves to obscure rather than to clarify. Capitalism is the prevailing economic formation in most countries. To say that it causes urbanisation is to advance explanation and understanding very little, as all social outcomes, both structural and spatial, are the products of capitalism. Such arguments have some validity at the most general level, but they fail to distinguish between capitalism as an underlying principle and as a specific and evolving economic formation. The value of independency theory lies not in its foundations in capitalism per se, but in the links that it proposes between successive stages in the evolution of capitalism and urban development across the world.

Stages of global urban development

The foundations for urban development in the core, and in localised areas in the periphery, were established up to about 1780 under conditions of mercantilism. This was an economic system that originated in the fifteenth century and involved the accumulation of wealth through trade. Its main feature was the buying and selling of the products of labour. These were principally agricultural and craft items. The aim of the merchant, as typified by Antonio in Shakespeare's *The Merchant of Venice*, was to buy goods at one price and sell them at a higher price, to consume some of the profits and reinvest the remainder in further trade. The highest profits could be obtained from long-distance trade in scarce commodities, so Antonio's ships were laden with exotic silks and spices from the Orient. Cloves were the most highly prized commodities in the mid sixteenth century, being more valuable than the equivalent weight in gold! Most trade followed established land or sea routes and took place with suppliers in well-defined source areas, so the Mediterranean, the Baltic, the north Atlantic, the Indian Ocean and central Asia emerged as distinctive trade areas. Buying, selling and consumption, however, were restricted to towns and cities, where sources of finance, trading opportunities and good communications were available. The control

Plate 3.1 *Venice in the seventeenth century was the crowning achievement of mercantilism*

points for mercantilism included many of the major cities of Renaissance Europe, of which Venice was by far the most successful. Under mercantilism, wealth generally originated in rural areas, whereas expenditure and consumption took place in urban centres (Johnston, 1980).

An important feature of mercantilism was the belief that the volume of trade was finite, so wealth could best be accumulated by capturing supplies and markets from rivals. This competition for territory and its products was the prime driving force behind exploration, discovery and colonisation in the sixteenth and seventeenth centuries. It was a process that was endorsed by national governments and was led by monarchs, aristocrats and privately-owned companies who sponsored exploration and settlement in the hope that it would lead to new trading opportunities and the creation of vast wealth. An example is the Virginia Companies of London and Plymouth, which were established in 1606 with the express purpose of settling the Chesapeake Bay area of North America so as to produce the wine, citrus fruits, silks and spices to compete with those that were the monopoly of Spanish merchants (Mitchell, 2001). In the event, the climate was found to be unsuitable and the principal crop was tobacco. The British and Dutch East India Companies and the Royal Niger Company are similar examples of early colonial organisations that were set up to promote settlement and to develop trade.

Table 3.5 *The world's ten largest cities in descending order of size, 1550–2000*

1550	1700	1900	2000
Paris	London	London	Tokyo
Naples	Paris	New York	Mexico City
Venice	Lisbon	Paris	São Paulo
Lyon	Amsterdam	Berlin	Mumbai
Granada	Rome	Chicago	New York
Seville	Madrid	Philadelphia	Lagos
Milan	Naples	Tokyo	Los Angeles
Lisbon	Venice	Vienna	Calcutta
London	Milan	St Petersburg	Shanghai
Antwerp	Palermo	Manchester	Delhi

Source: updated from Chase-Dunn (1985).

Mercantilism was responsible for establishing the foundations for urban development in colonial powers. It made possible the introduction of highly profitable trading links that led to the generation and concentration of wealth in cities. In 1500 the world's ten largest cities were all in Europe (Table 3.5). At the other end of the trade route, there was only limited and localised urban development, as the role of the periphery was to supply, but not to process, basic agricultural products and raw materials.

The process of colonisation extended over many decades and therefore led to the creation of urban patterns of varying complexity. In some territories there were existing urban structures upon which colonial influences were introduced, but mostly there was no prior urban settlement of significance and so the pattern that developed was wholly colonial in character. The ways in which towns and cities were introduced into, and grew in, areas in which there was little or no existing urban development, are described and explained by Vance (1970) in his mercantile model of urban development that is based upon the east coast of North America. A modified version is shown in Figure 3.7. The model identifies a sequence of stages through which initial contact leads to differential urban development in the core and the periphery. For Vance, the first phase of mercantilism involved the exploration of overseas territory and the search for information about the potentials for production and trade. Once favourable reports were received, staple products such as fish, fur and timber were harvested, but no permanent settlement was established (Stage 2). Urban development began in Stage 3 with the first settlement of colonists after 1620, who both produced staples and consumed products manufactured in the home country. The two-way trading links that were established focused upon the principal port that became the administrative centre of the colony.

The subsequent stages trace the development of the urban system under capitalism. The fourth stage is distinguished by the development of internal trade and manufacturing in the colony and the extension of transport links and trading routes from the major gateways into the interior. At the same time, there is rapid growth in manufacturing in the homeland to supply both the overseas and the domestic market. The final stage is achieved when a fully-fledged mercantile settlement pattern exists in the colony and is matched by a mature industrial system of cities, organised on a central place basis, in the imperial power.

An important characteristic of the settlement pattern in the colony is its linearity. Settlements are aligned along the coasts and are also located

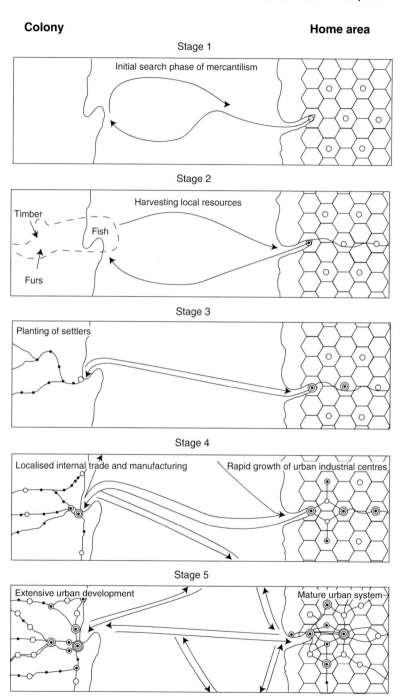

Figure 3.7 *The mercantile model of urban development*
Source: adapted from Vance, 1970.

along the routes of trade that connect the coastal points of attachment to the staple producing interiors. Over time these become integrated within a functionally interdependent system of cities (Friedmann, 1972). Close historical and geographical parallels exist between the patterns of urban development in the USA, as modelled by Vance, and those in western Africa and Brazil, as identified by Taaffe, Morrill and Gould (1963). The work undertaken by these authors suggests that the principal contribution of mercantilism was the creation of a well-developed urban pattern in the core, but an embryonic urban framework, except at points of contact where exporting centres flourished, in the periphery.

Urbanisation in a capitalist world economy

The spatial patterns of urban development that were established under mercantilism and early colonisation were accentuated and compounded when capitalism and imperialism became the dominant economic and political systems in the late eighteenth century. Capitalism is a form of economic organisation in which wealth is generated for investors through the production of saleable goods and services. Its main feature is that the capitalist employs workers directly rather than merely bargaining for and trading in the items they produce. Profits are made from the differences between the value of the product of labour and the price that is paid for it. To be most successful it requires large inputs of raw materials and extensive markets, which are best ensured through the possession of empire. Capitalism, through mass production and associated agglomeration, generates urban growth and urbanisation because it concentrates productive activity and all the workers and spending power that are associated with it. In addition, the city continues to serve as a centre for the consumption of the profits of capitalism. Under mercantilism, cities are the points of consumption and the articulation of trade. Under capitalism there is a third function, organised mass production, which is linked to the other two (Johnston, 1980). At the same time, the increased volume of trade stimulates additional urbanisation in dependent overseas territories.

With the benefit of hindsight it is possible to trace the evolution of capitalism through distinct industrial, monopoly and transnational corporate stages since it first became the dominant economic system in late-eighteenth-century Britain (Table 3.4). The individual stages are distinguished by both structure and space relations. The former relates to the underlying economic formation and is reflected in the representative type and unit of production. The latter refers to the geographical locations and areas in which wealth accumulation principally occurred. Each stage gave rise to, and accelerated, urban development in different parts of the core and the periphery.

The initial phase was industrial capitalism in which wealth was created by making rather than merely trading in goods. Manufacturing involved large numbers of people who tended machines that performed sets of routine and repetitive operations in order to make standard products. Large inputs of raw materials were involved, so industrial capitalism was supported through extensive trade with overseas suppliers of fibres, ores and agricultural products, many of whom were controlled through colonial administration. Although small by today's standards, the early industrial factory employed many more people than the craft workshop that it succeeded. As such, as in Great Britain, it generated sizeable concentrations of population and labour that in turn attracted more industry and led to rapid and massive urban growth (Box 3.5). Increased

Box 3.5 Urban development in industrial Britain

Great Britain was the first country to experience urban growth and urbanisation as a result of industrialisation. The industrial revolution, which began in the last third of the eighteenth century, transformed the country from a rural agricultural to an urban industrial economy in less than 100 years. The pace of population growth was unprecedented and unparalleled. At the first census in 1801 the total population of England and Wales was some 8.9 million. By 1891, the last census of the century, it had risen to 29 million, an increase of 326 per cent. The urban population grew by 946 per cent. Between 1801 and 1851 cities accounted for two-thirds of the population increase of 9 million people in England and Wales. A population that was 26 per cent urban in 1801 was 45 per cent urban in 1851. By 1861, for the first time in any country, more people in England and Wales lived in towns and cities than lived in rural areas.

Industrial capitalism created a new pattern of urban settlement in Britain termed by one contemporary observer 'the age of great cities'. At the beginning of the nineteenth century, London, with some 861,000 people, was the largest city in the world, exceeding Constantinople (570,000) and Paris (547,000), but it was the only place in Great Britain with over 100,000 people. By 1851 its population had risen to 2.4 million and there were two other British cities, Liverpool and Manchester, with over 300,000. Birmingham, Leeds, Bristol, Sheffield and Bradford had between 100,000 and 300,000 and there were a further 53 cities between 10,000 and 100,000 in size.

demand was similarly translated into urban development in economically linked parts of the periphery.

In contrast to its profound impact upon settlement patterns in Great Britain, industrial capitalism did little to change the urban/rural balance of population elsewhere. Although industrialisation had spread to adjacent parts of north-western Europe and to North America by 1890, the level of urban development in the core remained low. It was negligible, when measured at the national scale, in the periphery. It is this urban world which was analysed and illustrated in detail by Weber in his classic work, *The Growth of Cities in the Nineteenth Century* (1899). The limited extent of urbanisation at the time is apparent when his data are mapped as far as is possible according to the present network of state boundaries. It goes without saying that this is a task of very crude approximation, which can only be undertaken with considerable cartographic licence, because the political geography of the world has changed so much over the past century (Figure 3.8). Only six countries – England and Wales combined, Scotland, Belgium, Saxony, the Netherlands and Prussia – were more than 20 per cent urban in 1890. With less than 3 per cent of the world's population living in towns and cities, there was little or no urban development in most other territories.

Such localised urbanisation as was produced in the periphery was, in accordance with Vance's model, a consequence of the concentration of population around points of supply (Figure 3.7). The industries of the core used domestic coal and iron ore to build and power machines to process cotton, sugar, jute, rubber, tobacco, wheat, tea and rice imported from colonies and imperial territories. These commodities were the products of agriculture and their primary accumulation in the periphery led to limited urban development, associated with shipment rather than with local processing. 'Sao Paulo grew on the basis of coffee, Accra on cocoa, Calcutta on jute, cotton and textiles, and Buenos Aires on mutton, wool and cereals' (Gilbert and Gugler, 1992: 47). Urban development in association with peripheral supply similarly took place in the West Indies and Indonesia, Malaysia and the Far East. Although cities were established along the coasts of empire, these developments did little to change the overwhelmingly rural distribution of the local population.

The reason for the limited impact was that the towns and cities that were established under colonialism were more closely linked to the urban system of the European power than they were to settlements in the

Figure 3.8 *Percentage of population urban, 1890*

Percentage of
population urban

☐ 0–9.9
☐ 10–24.9
☐ 25–49.9
■ > 50.0

surrounding area. The purpose was to facilitate economic imperialism rather than to service or promote economic development in the colony. The process of urban formation within the British Empire is documented in detail by King (1990), and the appearances of such towns and cities owed more to European styles than to vernacular forms. In 1800 the principal colonial cities were Calcutta, Bombay, Madras, Dacca, Sydney, Halifax, Montreal, Toronto, Port of Spain, Bridgetown, Gibraltar, Kingston and Nassau. A century later, by 1990, this had expanded to include Aden, Hong Kong, Cape Town, East London, Durban, Pretoria, Johannesburg, Salisbury, Blantyre, Mombasa, Kampala, Zanzibar, Lagos, Accra, Nicosia, Suez, Port Louis, Mahe, Kuching, Georgetown, Melbourne, Brisbane, Adelaide, Perth, Hobart, Christchurch, Wellington, Port Moresby and Port Stanley.

An important feature of these cities, emphasising their imperial connections, is that their built and spatial environments had much in common with each other, being western in inspiration, but were different to urban centres in the interiors of the countries in which they were situated (King, 1990: 140). Colonial cities had a very distinctive role as

Plate 3.2 *Western architectural styles in a colonial city: the railway station in Chennai (Madras), India*

Table 3.6 *Principal characteristics of colonial cities*

Geopolitical

1   External origins and orientation

Functional

2   Centre of colonial administration

3   Presence of banks, agency houses and insurance companies

4   Focus of communications network

5   Warehousing/distribution centre

Economic

6   Dual economy, dominated by foreigners

7   Presence of large numbers of indigenous migrant workers

8   Municipal spending biased towards colonial elite

9   Dominance of tertiary sector

10 Parasitic relations with indigenous rural sector

Political

11 Eventual formation of indigenous bureaucratic-nationalist elite

12 Indirect rule through leaders of various communities

13 Social polarity between superordinate expatriates and subordinate indigenes

14 Caste-like nature of urban society

15 Heterogeneous society comprising colonial elite, in-migrants from other colonial territories, in-migrant educated indigenes and uneducated indigenes

16 Occupational stratification by ethnic groups

17 Pluralistic institutional structure

18 Residential segregation by race

Physical/spatial

19 Coastal or riverine site

20 Establishment at site of existing settlement

21 Gridiron street plan

22 Presence of elements of western urban design

23 Residential segregation between colonials and indigenes

24 Large differences in population density between residential areas of colonials and indigenes

25 Tripartite division between indigenous, colonial and military cities

Source: modified from King (1990: 17–19).

nerve centres of colonial exploitation. They were places where the banks, agency houses, trading companies and shipping lines, through which capitalism maintained its control over the local economy, were situated. As such they had a number of well-defined physical and geographical features (Table 3.6). They were also the early peripheral links in the emerging world economy. Because of their early lead in urban growth, many subsequently became primate centres and, in some cases the capitals of independent states.

Monopoly capitalism replaced industrial capitalism and colonialism towards the end of the nineteenth century. It involved the ruthless exploitation of peripheral areas and was distinguished by a vastly increased scale of economic activity and the domination of newly created international markets, within state-controlled empires, by a small number of producers in each sector. Monopoly capitalism emerged in response to the demand for products that was generated by the rapidly growing population of the industrial nations. This stimulated manufacturers to diversify from making heavy, crude products into the mass production of a wide range of consumer goods and services. Increased output occurred both because the core economies in Europe became more productive, and because the manufacturing belt of the USA attained core status alongside the UK, France, Germany and the Low Countries during the 1880s (Chase Dunn, 1989). It was achieved through the consolidation of many factory enterprises into multinational corporations that typically engaged in many functions in many areas, both at home and in the periphery.

Monopoly capitalism involved the more ruthless exploitation of peripheral areas. The larger scale of industrial activity required the international sourcing of raw materials and the international marketing of manufactured products, so the success of the core regions became dependent on their ability to dominate and control overseas territories. This was either through formal imperialism, of which the British and the French empires were the largest in the early twentieth centuries, or else through corporate power and influence as increasingly exercised, for example, by the industries of the USA. Britain established itself as the leading imperial power after about 1880, when it increasingly drew its industrial raw materials, including ores, oil and rubber, from around the world and in return supplied its overseas possessions in India, Africa and the Far East, and other territories, with railways, ships, machinery, arms and motor vehicles. Similarly, the USA rapidly became a major international player after 1909, when, symbolically,

Selfridge's store was opened in Oxford Street London, at the very centre of the dominant power in the world economy (King, 1990: 81). Thereafter, many of the major US corporations, including Goodyear, Standard Oil, Ford and General Motors, developed international spheres of operation.

Monopoly capitalism produced further urban growth and urbanisation in an expanded core, although urban development in the periphery remained limited (Breeze, 1972). Precise comparison of the urban world in 1890 (Figure 3.8) with that in 1950 (Figure 3.6) is inappropriate because Weber's data for the nineteenth century are far less reliable and refer to a very different geopolitical era. The overall pattern of change, however, is clear. Urban development in the first half of the twentieth century occurred most rapidly and extensively in Europe, the Americas and Australasia. Most of the rest of the world was unaffected.

Urban development in the core was associated with the growth of cities as specialised centres of international business, alongside their traditional roles as places of administration and consumption. Examples in the UK include Glasgow, Newcastle, Manchester, Leeds and Sheffield. Chicago, Detroit, Philadelphia, Cleveland and Buffalo performed similar roles in the USA. Land-use patterns within such cities were dominated by a downtown area that housed the commercial heart of the urban industrial economy. The central business district (CBD) was the location for divisional and group head offices servicing corporate empires, for higher order retailing which attracted customers from the furthest reaches of the urban market area, and for those in government and in retailing who provided municipal services and convenience goods for the city residents themselves. Intense competition between these different uses pushed rents to high levels, so the land value surface peaked and declined steeply with distance away from the centre of the CBD. The existence of a strong and prominent CBD with its concentration of skyscraper office buildings clustered tightly around the central zone of conflux, as in the case of Chicago (Plate 3.3), symbolised the overriding importance of physical accessibility in the industrial metropolis.

Much of the urban pattern in 1950 is explained by processes of population concentration that were associated with the economic and political imperialism of the UK, the USA, Russia and France. High levels of urban development in Canada, South and Central America were a legacy of British trade and, more recently, corporate links with the USA. Limited urban development existed across the Russian empire in Asia, and central and eastern Europe. Urbanisation elsewhere in the

Plate 3.3 *Urban form under monopoly capitalism. Chicago in 1970 was dominated by a central business district of offices, shops and civic amenities*

periphery was largely a product of British and French imperialism. Although only one-quarter of the population lived in urban places, the principal feature of the urban world in 1950 was that the cycle of urbanisation in the dominant economies of the core was, or was very nearly, complete. In the periphery it had hardly begun.

Conclusion

This chapter has examined the historical processes of urban growth and urbanisation at the global scale. The emphasis has been on the forces that were responsible for the creation of large cities and societies with predominantly urban populations in the developed and the developing worlds. Both urban growth and urbanisation have long histories and were accelerated under mercantilism, but the progression towards the contemporary urban world did not gain any significant and sustained momentum until the industrialisation, in the early nineteenth century, of some of what are now numbered among the core economies. This key development was a consequence of the emergence of industrial capitalism as a dominant economic and social formation and of its external relationships that were formalised through colonialism and imperialism. It led to major and rapid urban development in the core economies of north-western Europe and North America and, subsequently, through relationships of interdependency, established the foundations for urban development in selected locations in the periphery.

Urban growth and urbanisation were reinforced and extended by monopoly capitalism in the late nineteenth century. The urban development that resulted was largely restricted to the core areas and to the coasts of empire, so that the world in 1950 was highly differentiated in urban terms. The next chapter examines the progressive emergence of an urban world over the past 50 years, as urban growth and urbanisation spread, under corporate capitalism, throughout the periphery.

Recommended reading

Breeze, G. (1972) *The City in Newly Developing Countries: Readings in Urbanism and Urbanization*, London: Prentice Hall. A set of readings on urbanism and urbanisation in the first half of the twentieth century. It includes the seminal paper by Kingsley Davis on the urbanisation of the human population.

Hall, P. (1998) *Cities in Civilisation*, London: Weidenfield and Nicolson. A massive, masterly, but very readable account of the golden ages of 21 of the world's greatest cities. Especially relevant chapters are those on the rise of the industrial city of Manchester, 1760–1830, and of the trading centre of Glasgow, 1770–80.

Gilbert, A. and Gugler, J. (1992) *Cities, Poverty and Development*, Oxford: Oxford University Press. A comprehensive analysis of the causes, characteristics and consequences of urbanisation in Africa, Asia and Latin America. The chapter by Gilbert on urban development in a world system is especially useful.

King, A. D. (1990) *Urbanism, Colonialism and the World-Economy: Cultural and Spatial Foundations of the World Economic System*, London: Routledge. A detailed and comprehensive overview of the literature on the contribution of colonialism to urbanisation at the global scale. The book explores the links between the metropolitan core and the colonial periphery and assesses the contribution of colonialism in creating the urban characteristics of the two areas.

Taylor, P. J. (1993) *Political Geography: World-Economy, Nation-State and Locality*, London: Longman. A general political geography of the world economy that includes a highly detailed account and explanation of the rise and fall of colonies and world empires.

Topics for discussion

1 Evaluate critically the relative contributions of natural change and migration to contemporary urban growth in the major cities of either Africa or Asia.

2 Outline and account for the causes of urban growth and urbanisation in Africa and Asia in the first half of the twentieth century.

3 Identify and account for the form and function of a named colonial city.

4 'Urbanisation is an anachronistic concept in the context of modern western society'. Discuss.

5 Outline and account for the major changes in the urban geography of Britain in the first half of the nineteenth century.

6 'The urban geography of nineteenth century Britain was dictated by the requirements of industrial production'. Discuss.

4 Urban development as a global phenomenon

By the end of this chapter you should:

- understand the nature of recent changes in the organisation of the global economy;
- be familiar with the changes in the political map of the world which have occurred since the end of empire;
- understand the consequences for urban growth and urbanisation of recent global economic and political change;
- understand the nature of debates surrounding the implications for urban development of global restructuring.

Introduction

The world has recently become urban because of major changes in the distribution of population in developing countries. High levels of urban development existed in the core economies of Europe and North America in the middle of the last century because of successive and cumulative urbanisation under mercantilist, industrial and monopoly capitalism. In the 50 years since, massive urban growth and urbanisation have occurred throughout most of the rest of the world, transforming it, and in consequence the world as a whole, from a rural to an urban place. At the same time, the pace of urban development in the core has slackened appreciably. This chapter examines and attempts to account for urban development as a global phenomenon.

Urbanisation became a global phenomenon in the last half-century as a consequence of deep-seated and far-reaching changes in the structure and spatial relations of capitalism. Two principal developments were involved: the replacement of monopoly capitalism by transnational corporate capitalism, and the creation of patterns of production, trade and service provision, which, rather than being restricted to the north Atlantic, or to political empires, are truly global in extent (Table 3.4). The former represents the emergence of a mode of production and consumption that is suited to accumulation in a rapidly expanding world market. The latter is a response to the end of imperialism, and the achievement of political independence by many colonies, dependencies and satellite states. These developments in combination define a new world economic order, with distinctive structural and spatial characteristics, a principal consequence of which was and is rapid urbanisation in many of the world's peripheral areas.

The organisation of production and consumption under conditions of transnational corporate capitalism within a new world system of nations is producing urban development in the developing world for two main reasons. The first is because investment in manufacturing and services by global capitalists is concentrated in selected cities in the periphery, which, as centres of economic and social opportunity, have become points of growth. The second is because large numbers of workers are being displaced from the land and flock into towns and cities, as traditional subsistence farming is replaced by globally oriented commercial agriculture. Major cities, commonly national capitals, are principal beneficiaries of these processes, many, as a result, increasing in importance as primate centres.

It is important to emphasise the speed with which urban development has transformed patterns of settlement across the world since 1950, and the scale on which it has occurred. The changes are captured by maps that show levels of urbanisation in 1950, 1970 and 1990 (Figures 3.6, 4.1 and 4.2). Africa and Asia were almost wholly rural in 1950 and it is here that the subsequent transition to urban living is most marked and has had, and is having, the most profound consequences. Major urban development, measurable at the national scale, began to affect parts of Africa and the Middle East between 1950 and 1970, although many countries in these regions, especially in the Sahel belt of Africa and in south-western Africa, were largely unaffected. No country in Africa was more than 50 per cent urban in 1970.

Urbanisation began significantly to affect the countries of southern and eastern Asia much later, as most did not pass even the 20 per cent urban mark until 1990. This lag, involving the highly populated countries of China and India, as well as Indonesia, Bangladesh and Pakistan, meant that the level of urbanisation at the global scale remained low. The periphery was substantially more urbanised in 1990 than in 1950, although the pattern in Africa and Asia was highly varied. Average levels were well below those in South America and in the core.

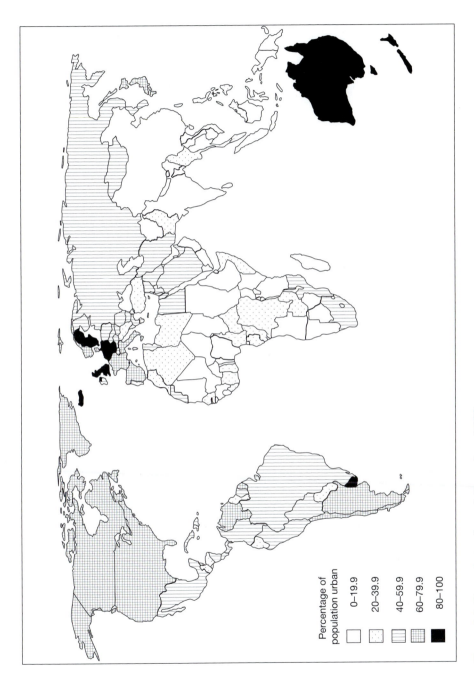

Figure 4.1 *Percentage of population urban, 1970*

Percentage of
population urban

0–19.9

20–39.9

40–59.9

60–79.9

80–100

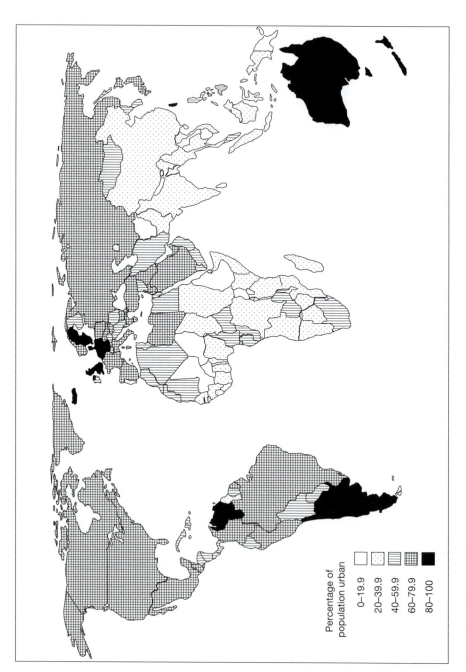

Figure 4.2 Percentage of population urban, 1990

Percentage of
population urban

0–19.9
20–39.9
40–59.9
60–79.9
80–100

There remained many countries in the remoter regions of Africa and Asia in 2000 with fewer than 40 per cent of their population living in urban places (Figure 2.3). Urban development has become a global phenomenon, although its full effects have yet to be felt in these regions.

The new global economic order

Urban development became a worldwide phenomenon over the last 30 years because of fundamental changes in the organisation and location of production and services as transnational corporate capitalism succeeded monopoly capitalism. A new economic order has emerged, characterised by global manufacture, and managed and controlled from the core economies by transnational corporations (Table 4.1). It is based upon a new international division of labour in which management, development and design take place in the core economies and routine production and service provision are located in the periphery. It is associated with the global organisation of business, finance and services. This new arrangement became possible because of, and owes much of its success to, the situation of relative peace and political stability at the global scale that has existed since the end of the Second World War.

The key change is in the location of manufacturing. Over the past half-century, an increasing volume of goods production has been arranged globally rather than within the narrow confines of nation states or empires, as was the case under monopoly capitalism. Much manufacturing has shifted from the core to the periphery, where the availability of very cheap labour enables standardised production to be undertaken at very low cost. Developing countries are estimated to have generated around 30 per cent of world manufacturing output in 2001, as opposed to around 12 per cent in 1970 (UNCTAD, 2002). The value added by industry in the developing world rose by an estimated 3.9 per cent per annum during the 1990s, as compared with 2.6 per cent in the developed world (World Bank, 2002). Much of the expansion in manufacturing occurred in the newly industrialising economies of China, Hong Kong, Brazil, Argentina, Mexico, South Korea and Singapore. Together these countries received 70 per cent of all foreign direct investment in 2000 (UNCTAD, 2002).

The types of production that have been shifted to the periphery include technologically more advanced products, such as pharmaceuticals,

Table 4.1 *Key features of the new global economic order*

Development	Agencies
Globalisation of manufacturing	Transnational corporations, newly industrialising countries
Globalisation of consumption	Brand management agencies Advertising agencies Marketing agencies Merchandising agencies
Globalisation of food supply	Supermarket chains, newly agriculturalising countries
Globalisation of finance	Banks, securities houses, exchange dealers, brokers, insurers
Globalisation of corporate services	Advanced producer service providers
Globalisation of personal services	Insurance companies, banks, travel agencies, hotel chains

Consequences for investment

Rise in foreign direct investment

Consequences for employment

Growth in the number of transnational producer service workers

New international division of labour

computers, scientific instruments, electronics and synthetic fibres; large-volume medium-technology consumer goods, such as motor vehicles, tyres, televisions and refrigerators; and mass-produced, branded consumer goods, such as cigarettes, soft drinks, toilet preparations and breakfast cereals (Dicken, 1998). The spatial patterns associated with the globalisation of manufacturing are illustrated by the geographies of the automobile and electronics industries. Production in both is scattered over a large number of countries in both the developed and developing worlds, but increasingly in the latter. Service functions, such as management, research and development and marketing, are almost exclusively restricted to the former (UNCTAD, 2002). Many of the firms that have shifted their production to the developing world have household names, are well established and are traditional linchpins of local economies in developed world countries. Such firms typically retain a token presence in their indigenous areas, but, following the shift, generate most employment and profits from their developing world operations (Box 4.1).

Parallel changes have occurred in agriculture as supply of an increasing range of farm products has been organised on a global basis. Exotic fruits, vegetables, flowers and poultry have joined the nineteenth-century staples of tea, coffee, cocoa, sugar, pineapples, citrus fruits and rubber on the list of agricultural products of the developing world that are grown on a commercial basis for supply to world, predominantly developed world, markets (Box 4.2). The volume of high-value foods in global circulation has increased especially rapidly as consumers in core economies have acquired a taste for ethnic cuisine. The value of world trade in such products accounted for 5 per cent of global commodity

Box 4.1 The new geography of manufacturing

A substantial amount of UK manufacturing has been shifted to the developing world over the last two decades and the trend is continuing. Shifts of production (all to southern Asia) announced in the first three months of 2002 alone include: Dyson washing machines from Wiltshire, Raleigh bicycles from Nottingham and Royal Doulton china and porcelain from Stoke-on-Trent. In the same period, Ford ended car production at Dagenham and Vauxhall (General Motors) at Luton and switched to imports. A transfer of production of Dr Martens, the footwear brand that epitomised the 1970's skinhead culture, to China from factories in Northamptonshire, Leicestershire and Somerset was announced in October 2002.

Box 4.2 The globalisation of agricultural production

Since 1990 an increasing number of African countries have entered the export trade in high-value foods and cut flowers (Barrett and Browne, 1996). One example is Kenya, which specialises in the export of high-quality, and often pre-packed, green beans, mangetouts, Asian vegetables (mainly okra, chillies and aubergines), avocados and cut flowers. Production is linked to European supermarket chains that oversee all aspects of production, exercise strict quality control and arrange for local packaging and bar-coding. They have developed highly efficient and fully integrated supply systems that ensure that products can be on supermarket shelves in a different continent within 48 hours of being harvested (Ilbery, 2001).

exchange in 1990 and increased by 8 per cent per annum throughout the 1980s (Goodman and Watts, 1997). The scale of changes involved, as traditional farming adjusts to meet the lucrative demand, is such that some of the economies concerned, including Brazil, Mexico, China, Argentina and Kenya, can be labelled as newly agriculturalising countries (Friedmann, 1993).

The transnationalisation of manufacturing involves the making of global products, with global brand names, which are assembled across the world from components made in a number of countries. It is commonly achieved by firms in the core economies investing directly in production facilities in developing countries – a practice that took off in the 1980s (UNCTAD, 2002). Firms that organise production on this basis are able to exercise control over the manufacturing process, while taking advantage of local conditions, especially the availability of large pools of very cheap labour. Ford's 'world car', which is designed, produced and built in strategically located plants around the world from parts fabricated in both developed and developing economies, is an example of a product which is made in this way. It is both manufactured and sold on a global basis. Such are the numbers of people who are incorporated within international systems of manufacturing that it is appropriate to talk of the emergence of a 'global factory' as the representative unit of production under transnational corporate capitalism.

An extreme form of global organisation is where all the production is subcontracted to developing world countries and the 'manufacturing' company concentrates solely upon management. Products are designed in the developed world, are manufactured in the developing world and are packaged and delivered direct to worldwide customers: the company has no direct involvement in making the goods that it sells (Box 4.3). Such firms are close to being 'virtual' in the sense that their major, indeed in some cases their only, asset is their brand name rather than their in-house production capacity. Their main activities are design, product placement, brand management and publicity.

Global production is principally undertaken by transnational corporations that have interests, affiliates and facilities in many countries. The United Nations Conference on Trade and Development recognises some 65,000 transnational corporations, together with more than 850,000 affiliated companies across the world (UNCTAD, 2002). In 2001 transnational corporations and their affiliates accounted for about 54 million employees, compared to 24 million in 1990, and

Box 4.3 Nike

Nike is an example of a virtual company that manages, but does not make, the products that it sells. Head office executives in Beaverton, Oregon are principally concerned with the design, advertising, marketing, merchandising and promoting of clothing and footwear, while manufacturing is undertaken by a complex network of subcontractors in South Korea, China, Indonesia and Taiwan. Different forms of subcontracting are used to ensure regular supplies of key products from reliable manufacturers, but to enable the company to dispense with the services of firms which fail to meet quality, price and production targets.

The organisation of Nike is almost the exact opposite of Ford in the 1920s and illustrates the extent to which the representative firm under global capitalism differs from that under monopoly capitalism. Ford at the time was the archetypal vertically integrated firm that owned and made money out of all aspects of car production, from the iron ore mine to the scrapyard. Large size and in-house control were the key requirements. Nike profits through brand management rather than manufacturing. Small size, flexible outsourcing and careful product placement are the secrets of its success.

generated sales of almost 419 trillion. They generate about one-tenth of world GDP and one-third of world exports. TNCs are important in all the major economic sectors, but are especially prominent in the telecommunications, petroleum, automobile, electronics, food, drugs and chemical industries.

Vodaphone is reckoned to be the largest non-financial transnational in terms of both foreign assets and the transnationality of its business as measured by an index based upon foreign assets, sales and employment. General Electric is ranked second, but is less involved in transnational business as it predominantly serves the US market (Table 4.2). Nestlé, with a transnationality index (TNI) of 95 per cent is reckoned to engage in the highest level of transnational activity of the top 20 non-financial TNCs, as it produces and/or markets food in most countries of the world. The global importance of transnational corporations is underlined by the fact that, in 1999, the average index of transnationality for the top 100 was 53 per cent. The largest 100 non-financial TNCs account for more than half of the total sales and employment of all the non-financial TNCs and this share is increasing rapidly. The headquarters of some 91 of the top 100 are in the USA, the EU and Japan.

Table 4.2 The largest non-financial transnational corporations, 2001

	Corporation	Home economy	Industry	Assets* Foreign	Assets* Total	Employment Foreign	Employment Total	TNI**
1	Vodafone	UK	Telecommunications	221,238	222,326	24,000	29,465	81
2	General Electric	USA	Electrical equipment	159,186	437,006	145,000	313,000	40
3	ExxonMobile	USA	Oil	101,726	149,000	64,000	97,900	66
4	Vivendi	France	Diversified	93,260	141,935	210,084	327,380	60
5	General Motors	USA	Motor vehicles	75,150	303,100	165,300	386,000	31
6	Shell	UK	Oil	74,807	122,498	54,337	95,365	57
7	BP	UK	Oil	57,451	75,173	88,300	107,200	77
8	Toyota	Japan	Motor vehicles	55,974	154,091	n/a	210,709	35
9	Telefronica	Spain	Telecommunications	55,968	67,084	71,292	148,707	54
10	Fiat	Italy	Motor vehicles	52,803	95,755	112,222	223,953	57
11	IBM	USA	Computers	43,139	86,349	170,000	316,303	53
12	Volkswagon	Germany	Motor vehicles	42,725	75,922	160,274	342,402	59
13	Chevron Texaco	USA	Oil	42,576	77,621	21,693	69,265	47
14	Hutchison Whampoa	Hong Kong China	Diversified	41,861	56,610	27,165	49,570	56
15	Suez	France	Power utilities	38,521	43,460	11,728	173,200	77
16	Daimler-Chrysler	Germany	Motor vehicles	n/a	187,087	83,468	416,501	24
17	News Corp.	USA	Media	36,108	39,279	24,500	33,800	85
18	Nestlé	Switzerland	Food and drink	35,289	39,954	218,122	224,541	95
19	Total/Fina/Elf	France	Oil	33,119	81,700	30,020	123,303	48
20	Repsol YPF	Spain	Oil	31,944	487,763	16,455	37,387	29

Source: based upon UNCTAD (2002: Table iv.1).

Notes:

* $US millions.

** the transnationality index (TNI) is calculated as the average of the following three ratios: foreign assets to total assets, foreign sales to total sales and foreign employment to total employment.

A new international division of labour, associated with the global organisation of production and service provision, is an integral part of the new global economic order. Its key feature is a concentration of mass labour tasks in developing countries, and of management, organisation and design work in the core economies of the developed world. It has succeeded and replaced the old international division of labour, which existed under industrial and monopoly capitalism, under which the periphery supplied basic extractive and agricultural products that were then processed in the core. The new international division of labour has been created by global producers in response to differences of labour costs and skill, education and training levels around the world. Very-low-cost workers in developing countries undertake labour-intensive tasks of manufacture, assembly and service provision under the direction of, and to specifications drawn up by, well-paid, highly educated technicians and managers based in the developed world (Box 4.4).

Box 4.4 The new international division of labour

The new pattern of organisation of labour can be seen in the semiconductor industry. Research and management functions, which require intellectual labour, are located in cities in core areas, where the quality of life is high and where the presence of large research universities provides links with theoretical science and a pool of highly qualified graduates. Examples include Boston, San Francisco, Los Angeles, Dallas and Phoenix. These are leading centres of university-based research and development and the headquarters of American transnationals in the computing and electronics sectors. The other processes, including mask-making, wafer fabrication, product assembly and testing are dispersed in low-cost areas throughout the periphery. Labour-intensive assembly of computers and electronics products is especially important in Hong Kong, Singapore, Malaysia, Taiwan and Korea.

A second example is provided by international publishing. The various tasks involved in producing a book tend now to be separated geographically. Editing and proof-reading, which are the most skilled operations, increasingly take place in the core economies, while labour-intensive typesetting and printing are devolved to the developing world.

The division of labour in some service activities is the latest to be restructured on an international basis. Much work in banking, insurance and the travel industry is of a routine and repetitive nature and is being subcontracted to telephone call centres in the developing world. An example is the opening of a centre by British Airways in India to handle ticketing for its customers worldwide. A particular attraction of India is that it has a large number of well-educated, very low cost, unemployed workers.

Global finance

The internationalisation of production is made possible by, and in turn gives rise to, a new pattern of international finance. A global system of supply and circulation has emerged in recent years in place of the bilateral funding arrangements, tied to trading blocs and dominated by governments that existed in the mid twentieth century. The new system, like the old, is directed and controlled by the economies of the core through a small number of world cities and helps to sustain their dominance, as is discussed further in Chapter 7. It differs substantially, however, in terms of the volume and nature of capital flows that are involved, and in the activities and places into which investment is directed.

The most important features of the global financial system are its size and spread. Total foreign direct investment was estimated as US$1.3 trillion in 2000 as compared to a mere US$94 billion in 1982 (UNCTAD, 2002). More than 50 countries had an investment stock of more than US$10 billion in 2000, compared with only 17 in 1985. The largest single element (around 33 per cent) consists of official development assistance in the form of grants and aid to promote economic growth and welfare. Such funds are used for physical and social improvements in both urban and rural areas in developing countries. Around 15 per cent, however, is in the form of direct investment by foreign firms in economic projects and initiatives. The principal sources are companies in the USA, the UK, Japan, Germany, Canada and the Netherlands, which together contribute about 80 per cent of foreign direct investment funds.

In the 1960s, foreign direct investment was concentrated in the exploitation of natural resources, particularly minerals and hydrocarbons that were exported and processed elsewhere. In the past decade there was a pronounced switch into manufacturing as transnational corporations established production facilities in the periphery. The pattern of destinations also changed, with a shift away from primary producing areas towards countries with emerging manufacturing economies. After 1986 foreign direct investment was channelled first into newly industrialising countries in Asia and then towards Association of Southeast Asian Nations countries and China. Subsequent waves of significant foreign direct investment extended to Sri Lanka, Turkey, Chile and Mexico, thus incorporating these countries within the global manufacturing economy (UNCTAD, 2002; Sit, 1993).

The global financial system is mediated by a number of institutions, including the International Monetary Fund and World Bank, which set policies, regulate credit and exchange rates, and arrange and provide multilateral aid for development. It is facilitated by the development of multinational organisations, such as the General Agreement on Tariffs and Trade, the Organisation for Economic Cooperation and Development and the North American Free Trade Agreement, which seek to promote and influence trade. The system is dominated by a small number of powerful banks that rank alongside transnational corporations as global institutions (Table 4.3). Thirteen of the largest 15 banks are based in the core economies and four are Japanese. The major banks of Europe and North America, however, handle more foreign business. It is made possible by the growth of the US dollar and Eurodollars as international currencies and media of exchange. The global financial

Table 4.3 *The 15 largest banks, 2002*

Bank	HQ location	Capital*	Assets*
Citigroup	New York	58,448	1,051,450
Bank of America	Charlotte, NC	41,972	621,764
Mizuho	Tokyo	40,498	1,178,285
JP Morgan Chase	New York	37,713	693,575
HSBC	London	35,074	696,381
Sumitomo Mitsui	Tokyo	29,952	840,281
Crédit Agricole	Paris	28,876	496,421
Mitsubishi	Tokyo	25,673	751,480
UFJ	Osaka	23,815	616,485
Industrial and Commercial Bank of China	Beijing	23,107	542,235
Bank of China	Beijing	22,085	406,150
Deutsche Bank	Frankfurt	21,859	809,220
Royal Bank of Scotland	Edinburgh	21,830	519,991
Bank One Corp.	Chicago	21,749	268,954
BNP Paribas	Paris	21,748	727,325

Source: *The Banker*, July 2002.

Note: *$US millions.

network operates through banking and capital markets, which work on a 24-hour basis, trading stocks, shares, futures and commodities. An important consequence is the growth of global investment practice through which capital, often in the forms of aid, is spread widely through the periphery.

Global services

Developments in production and finance are associated with, and are in part dependent upon, the growth of the international service economy. Service activities that were once domestically bound have reorganised on a transnational basis so as to serve the needs of businesses operating across the globe. This trend is reflected in the rise of the advanced producer services sector, which provides support services to industry. It includes insurance, accountancy, real estate, legal, advertising, research and development, public relations and management consultancy firms.

The change to global operation has been most marked in those service sectors in which the level of international activity was historically limited. One such field is accountancy, where six big firms – Arthur Anderson, Ernst and Young, Deloitte Touche Tohmatsu International, KPMG Peat Marwick, Coopers and Lybrand, and Pricewaterhouse – dominate the global market (Beaverstock, Smith and Taylor, 1999). A similar pattern applies in advertising, banking and legal services. Global business is further facilitated by means of the organisation of employee services, including hotel accommodation, car hire and personal finance, on an international basis.

The new political geography

The emergence of a new political map of the world enabled, and in turn was made possible by, the new economic order. The key feature was the ending of imperialism by the UK, France, Belgium and the Netherlands between 1950 and 1980; and by Russia in the late 1980s (Figure 4.3). The former resulted in the creation of new nation states in many parts of colonial Africa and Asia. The latter led to the restoration of full political independence to the countries of eastern Europe and the establishment of new countries in central and southern Asia. These developments added further changes to the political map that had been transformed during

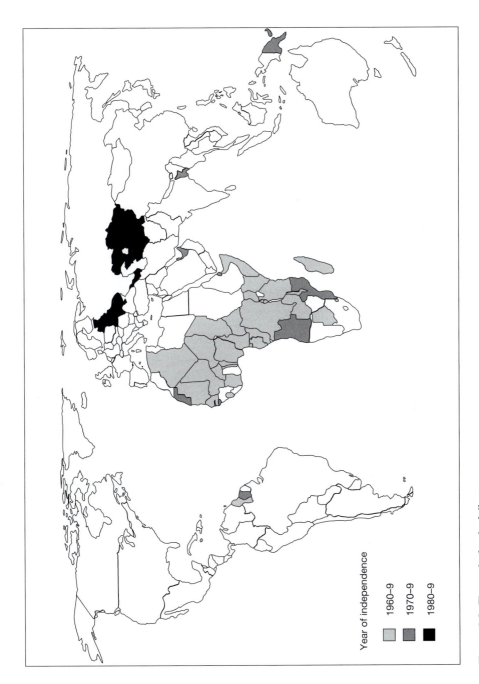

Figure 4.3 *The end of colonialism*

Year of independence

1960–9

1970–9

1980–9

the 1940s by the post-war redrawing of boundaries in Europe and by the withdrawal of the British from the Indian subcontinent.

A vast extension of capitalism, at the expense of alternative economic systems, accompanied the new political geography, further facilitating the process of globalisation. Imperialism effectively ended with the collapse of the Soviet Union in 1989. Communism is in retreat, although China, North Korea and Cuba maintain their centrally planned economies. The orientation of some of the former Soviet republics is presently unclear, but most seem keen to participate in the global capitalist economy so as to enjoy the benefits of trade and aid.

The new political pattern was created in conditions of relative peace and prosperity, certainly in comparison with those that prevailed in the first half of the twentieth century, with its two world wars and numerous regional conflicts. Global stability since 1945 principally arose out of the balance of power between the West and the communist bloc under which major wars were restricted to Korea and Vietnam. Localised disturbances and civil wars, often associated with decolonisation, were common among the newly independent states of Africa, Asia and eastern Europe, but, although some were bitter and protracted, most were constrained by national boundaries. The most important reasons were the political influence of the superpowers, especially the USA, and the mediating role of the United Nations. Terrorism, directed at symbols of global capitalism, as in the attack on the World Trade Center in New York in September 2001, has arguably replaced war as the greatest threat to world economic confidence.

Stability was further facilitated by the creation of supra-national and international organisations by many of the market economies, to undertake some of the traditional roles of the nation state. The main concerns were with defence, economic and social policy. Examples include the defensive alliance of nations in the North Atlantic Treaty Organization and the common market of economies in the European Union. The effect was to raise overall levels of international confidence and so create improved conditions for a restructuring of capital, for purposes of wealth accumulation, on a global basis.

The urbanisation of the developing world

The new economic order is principally responsible for the recent rapid urbanisation of the periphery, which in turn raised the level of urbanisation at the global scale beyond the 50 per cent mark. Since the mid twentieth century, and especially over the past 20 years, the global economy has subsumed local and regional economies across the world, so that most of the remaining peripheral countries and territories have been drawn into the world economic system. The role of the periphery, within the expanded world economy, is to produce low-cost manufactured goods and agricultural products and to act as markets for the products of the developed world. Transnational corporate capitalism produced and is producing urbanisation in the periphery both directly, as a consequence of urban growth in response to localised investment in manufacturing, and indirectly through its impact on subsistence and farming systems. The effect is to raise levels of urbanisation in South America, where there were already sizeable urban populations, and to initiate urbanisation across large parts of Africa and Asia, where, in 1950, it had barely started.

Urbanisation is triggered by investment because economic exchanges between core and periphery are spatially focused and so lead to a concentration of globally related economic activity in urban places. Cities, especially national capitals and those with major ports or international airports, offer significant advantages for developed world capital, affording wide access to cheap labour and to domestic markets. Such places are typically the major, and in some cases the only, centres in the country to have large-scale industry, hospitals, universities, media services, reliable Internet connections and facilities for sport and the arts. As cosmopolitan centres with good external connections, they are attractive to corporate managers and specialist workers on overseas postings. They are likely to be the home base of local elites who shape behaviour and consumption patterns towards which others in the country aspire.

The urban concentration of foreign investment-led economic activity is high across much of the periphery. In China, major coastal cities are the most favoured locations, such that 12 coastal regions accounted for an estimated 87 per cent of foreign direct investment in 1999 (UNCTAD, 2002). Guangdong alone held 29 per cent of all foreign direct investment stock in that year. In Thailand, three of the 68 provinces received one-third of foreign direct investment between 1987 and 2000. Overseas investment is similarly highly localised in Indonesia, being largely restricted to the area around Jakarta, where all major foreign corporations have their headquarters (Forbes and Thrift, 1987). Abidjan, the capital of the Ivory Coast, has 15 per cent of the national population, but accounts for more than 70 per cent of all economic and commercial transactions in the country. Bangkok accounts for 86 per cent of gross national product in banking, insurance and real estate, and 74 per cent of manufacturing, but has only 13 per cent of Thailand's population. Lagos, with 5 per cent of Nigeria's population, accounts for 57 per cent of total value added in manufacturing and has 40 per cent of the nation's highly skilled labour (Kasarda and Parnell, 1993: ix).

A strong metropolitan concentration of foreign investment is also noted in Latin America. São Paulo, with about 10 per cent of Brazil's population, contributes over 40 per cent of industrial value added and a quarter of net national product. It contains the production facilities of large transnational corporations and functions as a mediating point through which the domestic economy is integrated within the international market. Santiago has 56 per cent of Chile's manufacturing employment and contributes 38 per cent of national industrial output. The presence of such concentrated manufacturing activity in countries with few alternative sources of employment to subsistence or semi-subsistence agriculture acts as a powerful attraction to immigrants, so urban growth, which is stimulated by investment, is compounded by rapid population increase. For example, Sit (1993) argues that the accelerated urbanisation that stems from such investment and concentration led directly to urban growth rates which were double that of the population growth rate in most Latin American countries during the 1980s.

Urban growth and urbanisation occur in other parts of the periphery by extension. The rate of change depends upon the degree of functional and spatial integration of the domestic urban hierarchy. In countries with few other urban centres, or where they are only weakly interconnected, the extent of diffusion is limited and so the international contact points grow at a disproportionate rate. This is especially the case in post-colonial Africa, where primate cities have continually outperformed other cities (Stewart, 1997). It is also a feature of some Asian countries, where growing international investment in manufacturing is reinforcing the position of large centres. In countries with better developed urban hierarchies, urbanisation is, or is becoming, a nationwide phenomenon. In contrast to the historical pattern in the core, the direct urbanisation of the periphery is being imposed from the outside rather than being generated from within. In this respect it differs only in scope and scale from the processes of peripheral urbanisation that are associated with preceding forms of mercantile, industrial and monopoly capitalism.

Urbanisation is also taking place as an indirect consequence of the impact of transnational corporate capitalism upon the economies of developing countries. The central argument here is that major structural adjustments are forced upon developing economies as the price, or penalty, for incorporation within the world economy. These lead to the displacement of large numbers of workers from traditional occupations, who flock into the towns and cities and so contribute to urban growth and urbanisation (Box 4.5). Peasant farmers are foremost among those in developing countries whose livelihoods are undermined by the drive for production of farm products which will generate foreign currency, both to help reduce national indebtedness and to enable governments to

Box 4.5 Urban development in Zimbabwe

Many of the urban consequences of the absorption into the global economy are exemplified by Zimbabwe, a country that attained formal sovereignty in 1980 after 15 years of unilaterally declared independence (Drakakis-Smith, 1992). The modern urban system in Zimbabwe emerged under settler colonialism to facilitate the export of various commodities and the import of consumer goods. Cities were dominated by the white minority in the country and blacks were prohibited unless they had a job and accommodation. In the countryside, some blacks worked for white farmers, but most were engaged in subsistence agriculture. The population was 17 per cent urban in 1970. The favouring of the white colonialists, however, meant that social and health care services were city-based and significant differences in standards of provision existed between urban and rural areas.

This basic pattern was transformed during the 1970s as a consequence of increased foreign investment and the opening up of external markets for the products of Zimbabwe's farms and factories. Urbanisation occurred through net in-migration to jobs in cities, as the manufacturing sector increased its contribution to the gross national product from 10 per cent in 1965 to 24 per cent in 1980 (Stoneman, 1979). At the same time, the mechanisation of many of the larger commercial farms, and their increase in size, generated a surplus of black labour in rural areas. Movement into the cities increased significantly after 1980, when the legislation restricting ownership and residence in cities was relaxed and removed. Many traditionally white areas of Zimbabwe's cities rapidly became black (Cumming, 1990). Urban growth was compounded when families were reunited and birth rates rose. Some 33 per cent of the population was thought to live in urban places in 2000 and the population of Greater Harare was in excess of 1.5 million. The recent rapid urbanisation in Zimbabwe, in common with many African and Asian countries, is a consequence of structural and associated spatial changes that are associated with the transformation of a rural subsistence into an urban-based and politically independent commercial economy that is incorporated within the global economic system.

acquire the symbols of statehood, such as grand presidential palaces and national airlines. Many have been displaced from their traditional lands and means of subsistence by the introduction of commercial agriculture, such as the switch to production of exotic fruits and out-of-season vegetables for developed world consumers (Barrett *et al.*, 1999). They include large numbers of the very poor, who have no alternative sources of support and must move to the cities to look for work. Some find jobs sorting and packing horticultural products for air freighting to Europe (Plate 4.1). Droughts and civil wars, especially in parts of Africa, have further undermined the viability of traditional farming, leading to increased rural to urban migration.

The policies of post-colonial governments stimulate urban growth by further enhancing the attractiveness of towns and cities at the expense of rural areas (Auty, 1995). One way is through the exaggerated bias of government expenditures on infrastructure and services in favour of urban areas. Another is the higher wage rates and better employment protection that exist in cities because urban workers are organised into trade unions. A third is the effect of trade tariffs on the price of goods,

Plate 4.1 *Kenya: packing roses for overnight shipment to Europe. Commercial agriculture has displaced subsistence farmers, who have flocked to cities, contributing to urban growth and urbanisation*

which discriminate most against low-income peasant consumers, while a fourth is the decline in the demand for locally produced staples as urban consumers develop a taste for imported food items. Such policies are creating 'backwash urbanisation' by destroying the vigour of rural areas and suffocating the cities with the burden of the human casualties this process creates. The implications are seen in the rapid growth and dire social and environmental conditions of many African cities, and of others throughout the developing world that are swamped by large numbers of in-migrants who are looking for work and welfare.

Work on migration patterns in China by Goldstein (1993), however, shows that urbanisation is taking place in parts of the periphery for reasons that are largely unconnected with the emergence of the world economy. With the exception of a small number of economic sectors, China largely functions outside the world system and yet it has undergone major and rapid urban development in recent years. Evidence from national population surveys analysed by Goldstein suggests that urban growth and urbanisation are principally products of migration, with three-quarters of all population movements being from rural to urban areas. This is despite the operation of a strict residential registration system that attempts to control internal migration. Because of the difficulties of re-registering, the official statistics exclude large numbers of migrants who further swell the populations of China's principal cities. The number of unofficial migrants in the 23 largest cities in 1990 was estimated to be around 10 million. It is thought to be in excess of one million in Beijing alone.

Urban growth and urbanisation in China are principally a consequence of domestic economic and social circumstances. The growth of population in the countryside, together with the economic reforms that were introduced after 1979, created a vast surplus of labour that could not be absorbed by the already overextended rural economy. Cities are powerful magnets for displaced peasants whose annual incomes depended on the vagaries of the weather and, until the 1980s, on income distribution decisions by the collective leadership of the communes. Moreover, economic liberalisation allows and indeed encourages farmers to market their products in cities. Social advantages associated with a movement to cities include the opportunity to escape from restrictions on marriage and size of family, and from the burdens of having to look after elderly parents. Urban development in China is largely occurring from within as a consequence of the adjustment from a rigid and traditionally repressive rural subsistence to a more open and liberal service-based urban economy.

Detailed evidence on the precise nature of the links between global
production and urban development in the periphery is presently
fragmentary, both because the relationship is new and because the data
on recent urban growth in the world's least developed and poorest
countries are lacking. The true scale and significance of urbanisation in
the periphery have only recently become apparent. The research that has
been undertaken points to the existence of a general relationship
between investment in manufacturing, as a consequence of global
restructuring, and urban growth, but with wide variations from country
to country. Taiwan, Singapore and Korea, where there is a clear
connection, and China, where urban development is largely a
consequence of rural changes associated with economic liberalisation,
perhaps represent the extremes. It is important also to distinguish within
the periphery between experiences in South America, where levels of
urban development are historically high, and Africa and Asia, where
they are low. Both types of area have been affected by the same shifts
of production activity associated with the emergence of the new
economic order, but with different consequences. The effects on cities
and urban systems in the former have largely been to consolidate
existing patterns by compounding growth in existing centres. In Africa
and Asia they have created and accelerated urban development where
little existed before.

Conclusion

This chapter has identified, and attempted to account for, the processes
responsible for global spread of urban development. It has focused
specifically on the recent urbanisation of the developing world, as it is
here that urban change is taking place most rapidly and is having the
most profound consequences: most of the developed world was
urbanised by the middle of last century. Settlement patterns across large
parts of the developing world have been transformed in recent years, as
external investments have created jobs in cities and as workers,
displaced from the land because of the switch from subsistence to
commercial agriculture, have migrated into urban areas. Such changes
are seen as consequences of the progressive incorporation of their
economies within the global corporate capitalist economy.

It is sometimes difficult for analysts who view the world from
universities in the historic cities of Britain and Europe to appreciate that
urban development elsewhere is so recent and is taking place on such a

vast scale. The contemporary urban world is essentially a product of processes of population concentration that began to work powerfully in the core economies as late as the early part of the twentieth century, and have only become truly worldwide in extent and effect in the last 20 years. Urbanisation, when viewed at the global scale, is a contemporary phenomenon that, in being driven forward by transnational corporate capitalism, owes comparatively little to history. The same generalisations apply to the location of the urban population. Urban settlements have existed since the days of ancient Egypt and Mesopotamia, but most of the world's major cities, and nearly all mega-cities, are effectively less than half a century old.

Urban growth and urbanisation have transformed the global pattern of settlement over the past half-century. A world that is today predominantly urban has replaced one in which most people, in 1950, lived in rural areas. The switch to urban from rural has far-reaching implications for the daily lives of billions of the world's peoples. It affects their work, residence, health and longevity. The major socio-economic consequences of global urbanisation are explored in the next chapter.

Recommended reading

Daniels, P., Bradshaw, M., Shaw, D. and Sidaway, J. (eds) (2001) *Human Geography: Issues for the 21st Century*, Harlow: Pearson. A set of 17 concise chapters on a wide range of themes in contemporary human geography. The four chapters on 'production, exchange and consumption' provide useful insights into the geography of global production, consumption and finance.

Dicken, P. (1998) *Global Shift: Transforming the World Economy*, London: Harper and Row. The standard textbook on the global economy, its evolution, structure and geography. It is especially impressive on the nature and role of transnational corporations, newly industrialising countries and the new international division of labour.

Dodds, K. (2000) *Geopolitics in a Changing World*, Harlow: Pearson. A clear and comprehensive introduction to the political geography of the contemporary world.

Drakakis-Smith, D. (2000) *Third World Cities*, London: Routledge. A concise text that examines the course of urbanisation and its effects upon developing world countries.

Gugler, J. (1996) *The Urban Transformation of the Developing World*, Oxford: Oxford University Press. A collection of eight invited contributions that examine the course of urbanisation in the major regions of the developing world up to 1990. There are especially useful chapters on urban development in India and China.

Knox, P. and Agnew, J. (1998) *The Geography of the World Economy*, London: Arnold. An undergraduate economic geography text that provides a comprehensive and detailed compendium of the history and geography of the world economy. Strong on world system approaches.

Martin, R. L. (ed.) (1999) *Money and the Space Economy*, Chichester: Wiley. A wide-ranging collection of essays on the geography of banking, financial centres, money and the local economy, and on money and the state.

Key web sites

www.un.org   This is the main site for the United Nations. It provides links, via 'economic and social development', to 'population', which gives access to the biennial reports on *World Urbanization Prospects*, and to 'human settlements', which gives access to the *Global Report on Human Settlements* and the *State of the World Cities* report.

www.unctad.org   The home site for the United Nations Conference on Trade and Development, which can also be reached from www.un.org. It provides access to the annual *Trade and Development Report* and the *World Investment Report*.

Topics for discussion

1 What trends are taking place in the pattern of urban development at the global scale? How can they be explained?

2 Explain and evaluate critically the meaning of the term 'new international division of labour'.

3 Assess the contribution of rural-to-urban migration to urban growth in the contemporary developing world.

4 Define and give good examples of (a) newly industrialising countries and (b) newly agriculturalising countries.

5 Explain the reasons why rapid urbanisation tends to arise when a country is absorbed into the world economy.

6 Assess the implications of the end of colonialism upon urban development in Africa.

5 ▸ Socio-economic consequences of global urban development

By the end of this chapter you should:

- **be aware of the principal social and economic consequences of global urban development;**
- **understand the nature, reasons for and value of informal work and shanty town living;**
- **be aware of the servicing and health problems of developing world cities;**
- **understand the nature of debates surrounding the ways of tackling the socio-economic problems of cities.**

Introduction

A wide range of profound socio-economic consequences is associated with the global shift in the distribution of population to urban from rural places. Urban growth is adding large annual increases to the population of towns and cities in many countries that are struggling to cope with its consequences. Major problems of employment, housing, service provision and health result. These are widespread across the developing world, but are most acute in those parts of Africa and Asia in which the scale and pace of contemporary urban change are greatest. The inability of many cities fully to accommodate the increase in population by providing work, housing and services points to a situation of overurbanisation which has important consequences for the quality of life and social stability.

Coping with the consequences of urbanisation is one of the biggest challenges facing the world today. It has generated wide-ranging debates about what can and should be done and by whom. One argument is that the developed nations have a humanitarian and a moral responsibility, arising out of their historical exploitation of the developing world, to intervene on a major scale and so should contribute assistance and aid to provide housing and basic infrastructure. They are wealthy enough to help in a substantial way and it is unacceptable for the people of the developed world to enjoy the fruits of affluence while human beings elsewhere live in poverty and deprivation.

An alternative point of view is that the urban problems in the developing world are matters for national governments and must be resolved domestically in line with local conventions and traditions. Urbanisation takes different forms in different places and it is for the countries of the developing world to work out their own solutions. Such views extend well beyond the narrow confines of urban geography and link to wider issues of independence, self-determination, global power and politics. They flavoured discussions at the United Nations Conferences on Human Settlements, held in Istanbul (HABITAT 2) in 1996 and Vancouver (HABITAT 1) in 1976, and on Sustainable Development, held in Johannesburg in 2002, which sought to build a global consensus on the ways forward for the urbanising world.

A significant achievement of these meetings was to agree on targets towards which city governments could aspire and on a methodology for measuring progress. A key indicator, devised for HABITAT 2, is the City Development Index, which measures average well-being and access to urban facilities by urban residents (Box 5.1). Pronounced differences exist in the level of development of cities: Stockholm and Melbourne score highly in contrast to Port Moresby and Lagos (Table 5.1). The Index is based upon five constituent measures, of which waste disposal varies most widely, this being a major reason for the differences in

Box 5.1 The City Development Index

The Index measures average well-being and access to urban facilities. It is based upon five sub-indices: city product, infrastructure, waste, education and health.

City product is analogous to gross domestic product per head and is based upon household incomes. *Infrastructure* is a composite measure of the number of water, sewerage, electricity and telephone connections. *Waste* is an index of the volume of water and solids treated. *Education* is defined in terms of literacy levels and school enrolments and *health* refers to life expectancy and child mortality. The quality of the data underpinning several of these indices is highly questionable, but the CDI represents the best single measure of the level of development of the world's cities.

health and life chances between people in developed and developing world cities.

Socio-economic consequences of urban development are difficult to deal with because urbanisation, as Chapter 4 has shown, is presently taking place most rapidly and on the largest scale in the world's poorest nations – those that have the least well-developed economies, societies, polities and infrastructures. This is the opposite of the nineteenth century, when rapid urbanisation in the advanced economies of western Europe and

Table 5.1 *Development indicators for selected cities, 1998*

City	CDI*	City product	Infra-structure	Waste	Health	Education
Stockholm	98	94	100	100	94	100
Melbourne	96	90	100	100	94	94
Singapore	95	92	100	100	93	89
Hong Kong	92	89	99	99	91	81
Moscow	90	81	99	87	84	99
Seoul	86	65	98	100	89	78
Rio de Janeiro	80	82	87	63	82	84
Sofia	80	71	94	59	86	86
Hanoi	74	60	72	90	86	86
Havana	71	68	75	50	81	85
Jakarta	69	66	57	47	80	96
Ulaanbaatar	68	53	58	90	71	67
Lahore	61	71	79	50	65	41
Colombo	58	47	69	45	86	45
Bangalore	58	51	82	31	75	49
Dhaka	48	56	55	28	65	49
Phnom Penh	44	40	33	27	47	70
Port Moresby	39	69	18	10	58	42
Lagos	29	42	30	2	44	29

Source: United Nations (2001b: 116).

Note: * the City Development Index (CDI) is a measure of average well-being and access to urban facilities by individuals.

North America was associated with economic growth and increasing prosperity. Many developing world countries are politically unstable or weak and lack strong bureaucracies that can initiate and impose appropriate planning solutions to the problems of their cities.

The causes, as Chapter 4 further emphasised, are also fundamentally different to those that were responsible for the urbanisation of the developed world, being largely a local consequence of engagement with, and participation in, the global economy. They are imposed from the outside in response to global processes, rather than being products of indigenous growth and change. Economic and social development lag well behind urbanisation in most developing world countries, so that many people in the city, and especially recent in-migrants, at best find marginal employment in informal economic activities, and accommodation in shanty towns in which there are wholly inadequate levels of service provision and far-reaching implications for human health.

The urban informal sector

The emergence of an unorganised, unregulated and unregistered informal sector in many urban economies is a widespread consequence of contemporary urbanisation. It is principally a feature of cities in the developing world. Its existence was recognised and documented in the early 1970s, when it was observed that massive additions to the urban labour force, through high net in-migration and natural increase, were not reflected in the official employment statistics. People were creating their own work and were generating sufficient income to support themselves and their families. A distinctive, alternative pattern of economic activity and exchange had developed, paralleling that in the formal economy and providing both benefits and disadvantages for its many participants (Table 5.2).

The informal sector encompasses a large number and a diverse variety of service and production activities that operate on an individual or family-owned basis and use labour-intensive and simple technology. It includes activities such as the selling of food and drink, hawking, letter writing, shoe cleaning, bottle and waste collecting, fortune telling, knife sharpening, snake charming and street entertainment; and work undertaken by mechanics, carpenters, small artisans, barbers and personal servants (Plate 5.1). It also covers activities such as prostitution, begging, drug peddling and scavenging, which are on the margins of social acceptability.

Table 5.2 *Key characteristics of formal and informal employment*

Formal	Informal
Difficult entry	Ease of entry
Modern	Traditional
Corporate ownership	Family/individual ownership
Capital-intensive	Labour-intensive
Profit-oriented	Subsistence-oriented
Imported technology/inputs	Indigenous technology/inputs
Large-scale	Small-scale
Protected markets (tariffs, quotas)	Unregulated markets
Skills acquired within formal schooling	Skills acquired outside formal schooling
Most workers protected by legislation/ social security	Few workers protected by legislation/social security

Source: based on ILO (1972: 6).

Plate 5.1 *Informal sector employment: motor tricycle repair on the streets of Chennai (Madras), India*

Informal sector work is illegal in that it fails to comply with bureaucratic rules, regulations and licensing requirements, but it excludes activities such drug peddling, robbery, extortion, protection and pickpocketing, which are part of the criminal economy (de Soto, 1989). Much informal sector work takes place on the street, but may also involve production tasks carried out in small workshops and factories. Evidence assembled by the United Nations (2002) for 21 developing world countries suggests that one-third to one-half of all output is generated by the informal sector. It is as high as 51 per cent in Gambia, 49 per cent in Indonesia, 40 per cent in Mali and 39 per cent in Zambia. The size and relative importance of the sector are expected to increase significantly as many more people are added to cities in the developing world than can find work in formal employment.

There are many positive aspects associated with the informal sector. The first and most obvious is that it provides jobs and income for large numbers of people who otherwise would have no means of economic support. It provides a safety net in countries that lack basic welfare services. Many informal sector workers are escapees from rural poverty and the sector gives them first access to an urban economy in which there are opportunities and potentials. It is a major reason for the lower level of poverty in urban rather than in rural areas. Work in the informal sector may be undertaken by women and children to the benefit of the family. Women find it especially difficult to get jobs in a formal sector that is generally dominated by men and can combine informal work with child-rearing. In some countries they comprise the majority of informal sector workers.

The informal sector is valued for its vitality and dynamism. Despite very low levels of turnover, some surpluses are generated that are especially valuable in developing countries, where there are shortages of capital. It is a sector in which jobs are created at low or, in some cases, no cost, where few skills may be required and where elementary instruction, training and experience can be gained. It provides an opportunity to initiate and to develop business ideas. The informal sector is more likely to develop appropriate technologies and make effective use of resources, and it plays an important role in the recycling of waste materials, many of which find their way into the industrial sector or provide basic commodities for the poor (Plate 5.2).

Flexibility is widely seen as an attribute of informal sector employment, workers being able to adjust their work periods to suit demand and to chase markets. The street, with its opportunities for social contacts,

Plate 5.2 *Recycling waste materials: selling second-hand clothing in Tunis*

spatial mobility and flexible working, may be a better workplace than the sweatshop factories of firms in the formal sector.

Against these advantages and benefits must be set the low standards of living that are supported, long hours of work, the implications for health, safety and hygiene, and the ease with which vulnerable groups can be exploited. Women and children are at high risk (Plate 5.3). The absence of formal regulation means that competition for prime locations may be acute, leading to protectionism, racketeering and violence. The link between the informal and formal economies is a cause for concern. Global corporations can circumvent obligations to undertake ethical trading by subcontracting some forms of manufacturing, such as the making of footballs and athletic footwear, to the informal sector, which may employ very low-paid and child labour (Box 5.2).

There is considerable debate among theoreticians and planners as to the contribution made by the informal sector and what, if anything, governments should do about it. Early hostility has been tempered by more positive views that emphasise the benefits of relieving unemployment, providing a gainful alternative to crime and providing an

Plate 5.3 *Child labour: a seven-year-old carves stones boxes for the tourist trade in Agra, India*

Box 5.2 Making footballs in Pakistan

Thousands of Pakistani children are employed in making footballs for export in the Punjab city of Sialkot. They begin on low-grade balls before graduating to international-class footballs. The basic materials are exported from Europe and the children punch and stitch the balls by hand. Most join the workforce before reaching ten years of age. Children provide very cheap labour and are considered ideal workers because of their small and delicate fingers. The small wages they earn by working 10–12 hours a day provide their family with only basic necessities.

Child labour is illegal in Pakistan, but the law cannot be enforced because of social and economic conditions. Families that have several children depend on their wages for survival. They could not afford to send their children to school even if education was free. Many children are in bonded labour because their parents are in debt to factory owners.

Source: *The Times*, 29 October 1997.

outlet for entrepreneurial talent (Torado, 2000). It is, in any case, so large and well established in most developing world cities that it cannot be closed down. The sector is commonly supported by populist local politicians who regard it as a necessary and inevitable response to excessive and unjust regulation created by government in the interests of powerful and dominant groups (de Soto, 1989). It is endorsed by many economists who regard it as a source of economic potential rather than as a poverty trap (Cubitt, 1995). Many global products, such as soft drinks, jeans, films and trainers, are sold through the informal sector, so it is valued highly by some transnational corporations.

The alternative is to see the informal sector in a less favourable light as being little more than a means of survival in the absence of formal sector opportunities and welfare provision. The majority of workers subsist in unhygienic, dirty and dangerous conditions, although a small minority generate reasonable incomes and achieve economic security and upward mobility. The informal sector in South American cities is best seen as a pyramid. At the top are successful informal sector enterprises that employ wage labour and tend to be relatively stable sources of income and employment. At the base are a large number of subsistence operations that could never conceivably be successful and long-term. There is considerable fluidity in between, with some units rising in response to favourable conditions in their particular niche in the market, while others move down because of adverse demand and competition. The overall picture is one of struggle to survive rather than of a sector full of entrepreneurial talent to be celebrated for its potential to create an economic miracle (Thomas, 1995: 130). The most appropriate role for government is to accept that it exists and will remain, but to act to curb its worst failings, transgressions and injustices.

The informal sector is the economic lifeline for perhaps one-and-a-half billion people in the developing world. It is far too large and well established in most cities to be dismantled. The opportunities which it affords for survival in the city are major reasons for in-migration. Its existence is both a consequence and a cause of the current rapid urbanisation across large parts of the periphery.

Shanty towns

Much of the rapidly growing urban population is accommodated in the slums and shanty towns that form highly distinctive features of the built

environments of most major cities in the developing world (Plate 5.4).
Shanty towns are self-erected and usually illegal developments that
represent a response to acute social need. Few cities can provide
sufficient housing to accommodate urban growth and in-migrants would
be unable to pay for it if it was available, so the population is forced to
build and live in makeshift dwellings (Box 5.3). The *favelas* of Rio de
Janeiro, the *pueblos jovenes* of Lima, the bustees of Calcutta and the
bidonvilles of Dakar are merely the best known and most well
documented of the squatter communities that are thought to
accommodate over one-third of the urban population in developing
countries (Torado, 2000: 294). Many of those who are unable to find
space or to accumulate the necessary resources and materials to
construct a shanty dwelling live in a temporary shelters on the street
(Plate 5.5). The least fortunate have no home and simply sleep in the
open on the pavement (Box 5.4). They include large numbers of street
children.

Few reliable estimates exist of the number of shanty town dwellers in
individual cities, either because no data on any form of housing are

Plate 5.4 *Shanty town development in Manila, Philippines*

Box 5.3 Shanty towns in Mumbai

There is a severe shortage of accommodation in the cramped island city of Mumbai. Little affordable housing is available and over half of the city's 11 million residents live in slums. These informal settlements have grown up on private land (about 50 per cent of the slum population), state- and city-owned land (about 25 per cent) and land owned by the federal government (24 per cent). This last category includes more than 30,000 families living in shacks that are 1.5 m to 30 m from the major suburban railway lines. Living conditions in these railway shanties are the worst in the city. In slums on private and state-owned land the city government has extended basic services such as water and sewerage. Land along the tracks, however, is the property of Indian Railways, which argues that providing services will encourage further settlement. Residents have no clean water, electricity, sewerage or garbage removal. They are open to extortion because they are occupying space illegally. There is the ever-present risk of being hit by trains that kill an average of three slum dwellers a day. Trains travelling through the railway slums must slow down from 50 km/hr to 15 km/hr, extending the daily commuting time for 4.5 million people and increasing tensions.

Plate 5.5 *Living on the street in Chennai (Madras), India*

Box 5.4 Life on the streets

Many people in the poorest countries, including children, live on the street. Some live in small shacks on pavements that use walls or fences around buildings for support. They are without water, sanitation, electricity or heating. Those even less fortunate sleep in drainpipes, abandoned cars or out in the open.

Most pavement dwellers look to live there on a temporary basis before moving into a shanty town, but many fail to achieve this transition. Studies of homeless people in Mumbai show that most have lived on the street since their arrival in the city, which could be up to 30 years previously (Patel, 1985). Half of pavement dwellers in these studies lived in huts of less than 5 sq. m. Most, however, had their names on electoral rolls and over half the children attended school. Three-quarters of the children had been born in municipal hospitals. Over one-third of all households had savings and about one-quarter had bank loans.

collected, or because such developments are a political embarrassment and so are not recognised in official statistics. Squatter households are recorded as comprising 34 per cent of the total in Ankara (Turkey), 48 per cent in Managua (Nicaragua) and 19 per cent in the Burkina Faso city of Bobo Dioulasso (United Nations, 2002). 'Marginal living quarters', a category that includes improvised and related forms of housing, are similarly reckoned to comprise 16 per cent of the housing stock in Cairo and 6 per cent in Buenos Aires. The paucity of information on shanty towns severely limits the scope for informed debate on their role and future.

Shanty towns have many features in common. They typically consist of closely packed constructions of mud, thatch, timber or corrugated iron that are erected on any available space without planning permission or building regulations approval (Box 5.5). Many are built on government-owned land from which eviction is politically impossible. Residential densities are extremely high and drainage, sanitation and water supply are lacking or are deficient. Many developments are alongside factories, railway yards and highways, which present opportunities for casual employment, but are sources of danger, noise and pollution. Most occupants are poor and work in the informal sector.

The speed of development and ubiquitous nature of shanty towns mean that they are the focus of wide interest from analysts, planners and politicians. Attitudes vary from outright opposition to benign tolerance.

Box 5.5 The building of shanty towns

The example of San Martin, Buenos Aires, illustrates the way in which shanty towns are created, according to Hardoy and Satterthwaite (1989: 12). The research that they report describes in detail the activities of September 1981, when a small well-organised group of squatters invaded and occupied 211 hectares of abandoned private land. As word spread, some 3,000 people entered and established themselves in the settlement in a period of five days. Government efforts to bulldoze the area were resisted by women and children who stood in front of the machines, and attempts by the police to stop construction were thwarted when cordons were broken at night. At its peak the rate of building was staggering, with five or six houses being erected each day. By October 1982 the settlement was home to some 20,000 people. Although it was erected illegally, the development was planned, with space being left for access roads and community facilities. A democratic system of organisation and administration soon developed. Conditions in San Martin remained poor because the local authority refused to pave streets, install sewers or provide health care, so the residents were forced to make their own arrangements. By 1984 a health centre and a school had been built.

The cramped conditions, poor construction and the fact that they constitute a visual embarrassment to municipal authorities and high-income groups mean that they can easily be viewed in negative terms. Shanty towns house populations at very high densities, with increased risks to health, and commonly lack basic utilities and amenities. Houses are built to low standards out of materials that come to hand and provide inadequate protection from the elements. They are highly susceptible to fire and harbour disease-carrying rodents and insects. Many shanty towns are located in high-risk sites and are especially vulnerable to floods, landslides and the hazards associated with transport and industry. Their very existence draws attention to the failings of legal processes, planning powers and government policies.

Many observers, however, view shanty towns in more positive terms, since they can emerge as vibrant communities in their own right with strong forms of social organisation. Shanty towns are an important first destination for new arrivals to the city and help them to become familiar with, and assimilated within, urban life. They provide valuable accommodation for those who would otherwise be without shelter, and their construction makes considerable and creative reuse of materials. The residents constitute a sizeable market for those in informal sector business activities, and shanty towns provide a pool of cheap and accessible labour for urban industries. They are well established across the cities of the developing world and will not go away.

For these reasons, many governments have adopted a pragmatic approach and have shifted away from policies of opposing and demolishing shanty towns. Instead they seek to provide basic services and to help residents to upgrade their housing. The critical step in this process is that of empowerment, granting legal rights to land and enabling occupants to participate fully and openly in the economic and political life of the city. An example was the granting of legal rights to the residents of the Cidade de Deus (City of God) in Rio de Janeiro in January 2003 (Box 5.6). There are also cases, as in Mumbai, where city governments have acted to remove shanty towns, rehousing railway dwellers on edge-of-city sites. Politicians need to be responsive, as shanty towns house large numbers of people and their votes are an important factor in municipal and national elections.

Despite their present importance, shanty towns are recent additions to the cities of the developing world. The earliest is reckoned to be the

Box 5.6 The City of God becomes legal

In January 2003 the President of Brazil granted the residents of the City of God (Cidade de Deus) the rights to occupy their illegally acquired homes. The shanty city lies on the banks of the polluted Rio Grande river in southern Rio de Janeiro. It is home to 120,000 people, many of whom took part in building the city when they moved to Rio to find jobs in the economic boom of the late 1970s.

One likely beneficiary is unemployed electrician, Severino da Costa, who lives with his wife and four children in a rickety brick and corrugated shack. He says:

> At last we will get the deeds to our home. When we came here we just took the land because there was nowhere else to go. I have worked all my life to buy the construction materials to build it. It's all we have and I can't believe it could finally become mine.

The decree is at the heart of a pledge by the President to tackle poverty and wealth distribution. More than 6.6 million Brazilians are homeless and at least 40 million live in illegal *favelas*. Some 80 per cent of Brazil's wealth is in the hands of 10 per cent of the population. Legalisation, however, means that approval will be given to unsafe houses and slum development. The decision has also been criticised because the shanty is the base for gangs who control Rio's highly lucrative cocaine trade.

Source: *The Times*, 27 January 2003.

favela that was built in Rio in 1910, but the vast majority are products of the major and rapid urban growth and urbanisation of the population that has occurred since the mid twentieth century (Torado, 2000). Shanty towns may look unattractive and be condemned by many governments, but they fulfil a vital social need. They symbolise the vigour and spirit of self-help, born out of necessity, that lie behind the rapidly expanding cities of the developing world. Shanty towns evolve through a sequence of illegal occupation and building, initial antagonism and opposition from the municipal authority, official acceptance and recognition, and, finally, formal incorporation into the wider city. This mechanism is the principal way in which urban growth is occurring in developing countries. The demand for housing is such that shanty towns are likely to remain conspicuous features of developing world cities for the foreseeable future.

Service provision

Explosive urban growth has created enormous difficulties for municipal authorities in providing basic public services. It has imposed additional strains on infrastructures that in many cities in the developing world have long been deficient. In recent years, urban populations across the developing world have increased far more rapidly than service provision, with the result that many people have inadequate access to water supplies and sanitation. Some 87 per cent of the urban population of developing countries had access to safe water, and 72 per cent had access to sanitation, according to The United Nations Development Programme's *Human Development Report* (UNDP, 2002). The same source also suggested that 63 per cent of the urban population of sub-Saharan Africa had safe water and that 56 per cent had provision for sanitation.

There is strong evidence, however, that official statistics greatly overstate the extent and quality of provision and that the situation on the ground is far worse (United Nations, 2002). Every household in Accra is purportedly 'served' by piped water, but this means that there is a standpipe within walking distance, which could be as far as 100 m, and the system is often not operational. In Dar es Salaam, half of all households have piped water, but the service is erratic. Trunk and distribution losses amount to about 60 per cent of pumped water and an estimated half of that which reaches taps is lost through leaks and spills, resulting in a general delivery shortage. Because of the speed at which the city is expanding, pipes are commonly installed in narrow trenches after the development of housing, so they are easily disturbed by subsequent road construction and use. Many house connections are performed illegally with inappropriate materials, increasing the risk of contamination of supplies. Less than a third of the population in Jakarta has direct connections to a piped water system. Around 30 per cent depends solely on water vendors whose prices are up to 50 times those paid by households served by the municipal water company. The remainder relies on shallow wells, many of which are contaminated, deep wells or river water (United Nations, 2002). The situation in Dakar is worse, as only 28 per cent of households have private water connections, while 68 per cent rely on public standpipes and 4 per cent buy from water carriers (Hardoy *et al.*, 1992). Many areas of developing world cities are supplied with water from tankers that provide an irregular and unreliable service (Plate 5.6).

The provision of sewage services is similarly far lower than the official statistics suggest. Hardoy *et al.* (1992) estimate that around two-thirds of the urban population in the developing world have no hygienic means of disposing of sewage and an even greater number lack an adequate

Plate 5.6 *Tanker supply in the absence of piped water: Chennai (Madras), India*

means of disposing of waste water. Most cities in Africa and many in Asia have only a limited network of sewers, so human waste and waste water end up untreated in canals, rivers and ditches. An example is Dar es Salaam, where only between 5 and 10 per cent of the population is served (United Nations, 2001a). The system is very old and some sewers have collapsed due to lack of maintenance. About 75 per cent of the population uses on-site sanitation consisting of latrines, cesspits, septic tanks and soakaways. Where sewage systems exist they rarely serve more than the population that lives in the richer residential areas. Some 70 per cent of the population of Mexico City lives in housing served by sewers, but this leaves some three million people who do not. In Buenos Aires it is estimated that 6 million of the 11.3 million inhabitants are not connected to sewers.

Official data are misleading because of the use of inappropriate criteria to define 'adequate' and 'safe' and the fact that statisticians rarely challenge them openly. Organisations such as the World Health Organization, the United Nations and the World Bank are 'intergovernmental bodies', with management boards made up of representatives of member states, so it is difficult to question the validity of the data that governments report. To be 'adequate', water must be of good quality, readily available, piped to the house and affordable, but these criteria are often overlooked by governments so as to represent the situation in their cities most favourably. Thus among those classified as 'adequately served' are the inhabitants of settlements where hundreds of people have to share a standpipe, even if it is poorly maintained and delivers contaminated water. Households are still classified as served by piped systems even if the water is available intermittently or for as little as a few hours a week. In Mombasa there are many households with water pipes in their own homes that have seen no water in those pipes for years (United Nations, 2002). Understanding the shortcomings in official statistics is important because they inform, and in many cases confuse, debates about standards and rates of improvements in standards of living in the cities of the developing world.

Health

The implications for the health of urban populations of informal sector employment, poor housing and very low levels of public service provision are profound. Contaminated water and inadequate waste disposal create domestic environments that threaten health, especially in shanty houses occupied by the poor (Plate 5.7). They are major causes of morbidity and early mortality in many of the cities of the developing world today, as they were in the cities of north-western Europe in the nineteenth century (Box 5.7). Communicable diseases, a category that includes infectious and contagious diseases such as gastrointestinal and respiratory complaints, spread rapidly in large densely populated cities. They are essentially diseases of poverty, but their effects are selective. Children, the elderly and those who are malnourished are at greatest risk. Communicable diseases are the largest single reason for the high death rates and low life expectancy in the world's poorest countries (Table 5.3).

The cities of the developing world, however, are generally healthier than the surrounding rural areas despite the size and density of their populations and the rate at which they are growing. They have access to better drinking water and sanitation and are the points where medical and scientific techniques are first introduced and can reach most people at least cost. Fertility rates are lower, so there are fewer cases of complications of pregnancy. More births tend to take place in hospital,

Plate 5.7 *Unhealthy living: a shanty town adjacent to a highly polluted canal, Jakarta, Indonesia*

The nature of the hazards in people's homes and their implications for health are illustrated by a study of households in Accra, Ghana, reported by the United Nations (2001a: 109). It revealed that 46 per cent had no domestic water source, 48 per cent shared lavatories with more than ten other households, and 89 per cent had no home garbage collection. Cooking facilities in 76 per cent of houses consisted of wood or charcoal stoves, and flies were observed in 82 per cent of kitchens.

There is a strong relationship between the incidence of these hazards and child diarrhoea, the main cause of infant mortality in the city. Only 2 per cent of households facing two or fewer hazards reported diarrhoea incidents, whereas this figure was 39 per cent in households with three or four hazards. The incidence of factors is greatest in low-income households. Morbidity and mortality, especially among children, are consequences of poverty.

Table 5.3 *Extremes of life expectancy at birth, 2001*

Country	Life expectancy (years)
Japan	81.4
Switzerland	80.2
Sweden	80.0
Australia	80.0
Iceland	79.8
France	79.3
Italy	79.3
Austria	79.0
Swaziland	40.2
Lesotho	40.0
Botswana	39.1
Zambia	36.8
Zimbabwe	36.8
Malawi	36.3
Angola	36.1
Sierra Leone	34.2

Source: WHO (2002: 178).

so rates of perinatal mortality are lower. Communicable diseases can be more easily controlled in cities than in the countryside. Infant mortality rates in urban areas were less than those in rural areas in 18 of the 22 developing countries for which comparable statistics were reported by Gilbert and Gugler (1992: 67). The urban rate was more than 20 per 1,000 lower than the rural rate in Brazil, Ecuador, Ghana, Indonesia, Liberia, Mali, Morocco, Nigeria, Peru, Senegal, Thailand, Togo and Zimbabwe. Death rates are generally lower in cities because of the relative youthfulness of urban populations.

There is encouraging evidence that levels of health in many of the world's poorest cities are improving significantly. Death rates are falling, as a predominance of communicable diseases is being replaced by a preponderance of non-communicable diseases. The significance of this epidemiological transition is that diseases that are mass killers are succeeded by those that are person-specific. The rise of HIV/AIDS is a complicating factor and is a major threat to urban populations, especially in southern Africa (Plate 5.8). Urban populations are increasingly

Plate 5.8 *Health promotion in Lusaka, Zambia. AIDS is a major threat to the urban populations of southern Africa*

susceptible to diseases of affluence and those associated with social instability, and to violence, accidents and mental ill health. Addressing these health issues will be a key task for city governments in the future.

Conclusion

This chapter has examined some of the social and economic consequences of global urban development. It has concentrated upon Africa and Asia because this is where urban development is presently fastest and where the impacts are most pronounced. Informal sector work, shanty housing, low levels of servicing and poor health are among the most important products of urban development, but for a comprehensive picture it is also necessary to consider poverty, transport, governance and crime.

The rapidly growing cities of the developing world house many millions of people at extremely high densities, yet provide a range of opportunities and quality of life that are greater than those enjoyed in the surrounding rural area. They offer locations for services and facilities that require large population thresholds and sizeable markets to operate efficiently. They contribute disproportionately to national economic growth and social transformation by providing economies of scale and proximity that allow industry and commerce to flourish. Dire warnings of the imminent social or economic collapse of one or other of the cities of Africa and Asia appear periodically in the press, but such a disaster has yet to happen. Social conflict and economic catastrophes are normally played out across and between regions or nation states rather than exclusively within individual cities. Highly disturbing pictures of shanty housing, informal employment, inadequate services and poor health in Calcutta, Mumbai, Rio and Bangkok readily divert attention from what such places represent. Rather than focusing upon negative aspects, it is appropriate to see the major cities of the developing world as significant social achievements.

Recommended reading

Gugler, J. (1997) *Cities in the Developing World: Issues, Theory and Policy*, Oxford: Oxford University Press. A collection of papers on various aspects of developing world cities, including especially useful contributions on the urban informal sector.

Hardoy, J. E. and Satterthwaite, D. (1989) *Squatter Citizen*, London: Earthscan. A detailed examination of the lives and prospects of residents of squatter settlements, and informal sector workers in developing world cities.

Kasarda, J. D. and Parnell, A. M. (1993) *Third World Cities*, London: Sage. A useful overview of levels of contemporary urban development in the developing world, including detailed case studies of urban issues in selected countries.

Potter, R. B. and Lloyd-Evans, S. (1998) *The City in the Developing World*, Harlow: Longman. This introductory text examines the urbanisation of the developing world and its social and economic consequences. There are especially useful chapters on housing and employment.

Thomas, J. J. (1995) *Surviving in the City: The Urban Informal Sector in Latin America*, London: Pluto Press. A detailed analysis of the size, nature and contribution of the informal sector in the cities of Latin America. The role of the sector is debated and the implications for policy are addressed at length. Useful detail is provided in the form of case studies of informal sector work in selected cities.

United Nations (2001) *Cities in a Globalising World: Global Report on Human Settlements*, London: Earthscan. A comprehensive review of recent urban development and its consequences. There are especially useful sections on urban employment, health and services, and discussions of contemporary urban problems and policy requirements.

Key web sites

www.un.org   This is the main site for the United Nations and provides access, via 'economic and social development', to 'statistics' and 'human settlements'. The 'statistics' area gives access to a wide range of urban and rural statistics, including the millennium indicators and their constituent measures. The 'human settlements' link is a point of entry to the global urban indicators on selected cities, which are assembled and published by the UN's Global Urban Observatory (GUO).

www.who.int/en/   The home site for the World Health Organization. It provides on-line access to the annual *World Health Report*.

Topics for discussion

1 'Slums of hope': how valid is this description of the shanty towns in developing world cities?

2 Critically evaluate the social and economic contribution of informal sector employment in developing world cities.

3 What should urban governments do about shanty towns? Illustrate your arguments with reference to particular cities.

4 Explain the concept of the 'epidemiological transition' and discuss its value in understanding the changing nature of morbidity and mortality in developing world cities.

5 Outline and evaluate the differences in the health of urban and rural populations in the developing world.

6 ► Urban culture and global urban society

By the end of this chapter you should:

● have a broad understanding of the nature and meaning of urbanism;
● be aware of the effects of time and space on human interaction;
● understand the ways in which urban cultures are constructed;
● appreciate the nature of debates surrounding the nature, spread and impacts of urban culture.

Introduction

The progressive shift of people from rural into urban places is accompanied by profound and far-reaching changes in the ways in which many of them live their daily lives. Towns and cities are different from villages in physical, social and economic terms. They offer their residents a far wider range of options and opportunities and enable them to engage in many more interests and activities than are possible in rural areas. They are places with large numbers of people, factories, offices, shops and recreational facilities, which facilitate, support and promote a variety of lifestyles that are distinctively urban in character and differ fundamentally from those that occur elsewhere.

Traditionally, urban patterns of behaviour and identity were thought to be a simple function of place. They were restricted to and experienced by those who actually lived in the city. Today, urban influences are extended well beyond settlement boundaries by long-distance travel, telecommunications and the mass media. Many people in seemingly rural areas are exposed to urban attitudes and values. They have a wired identity and engage with the city via a range of symbols and shared meanings (Morley and Robins, 1995). The ability to participate in an urban way of life is largely independent of location and is open to all. The world, it is argued, is increasingly becoming a global urban society of which we are all residents.

'Urban' is a descriptive label which is used to describe both a particular type of place and a set of distinctive patterns of association, values and behaviour. It is this latter, sociological meaning, which is addressed and examined in this chapter. The concept of urbanism is that of a set of lifestyles and meanings, amounting to a distinctive urban culture, which arises in cities and follows from the impact of cities on society. It is expressed and reflected in patterns of social and economic relationship and behaviour, and through taste, fads, fashions, identities, aspirations and achievements.

The spread of urbanism is the least advanced of the principal processes of global urban change and is by far the most difficult to conceptualise, document, interpret and explain. Urban growth and urbanisation have created towns and cities which house a little over half of the world's population, but the progress and extension of urbanism, however defined, is much more limited. Some people who live in remote locations have lifestyles which are similar to those in cities. They participate fully in an urban culture and their attitudes and values are the same as those of residents of the world's principal cities. Most, however, do not. Equally, there is a large number of people, especially in the million and mega-cities of the developing world, who retain patterns of association and behaviour that are more akin to those in rural areas. Their attitudes and values are permeated and constrained by traditional influences of religion, the family and geographical parochialism. The lifestyles of many first- and second-generation in-migrants have not yet been affected by incorporation within urban society.

When viewed in this way, the spread of urbanism, like urbanisation, is seen to be a finite process. It begins with a wholly rural society in which urban influences are absent and ends when everyone everywhere lives an urban way of life. Mapping its advance is necessarily inexact since it involves evaluating changes in attitudes and behaviour which are not capable of precise specification and measurement. Cross-national data on urban growth and urbanisation are available, though of limited quality, as the Appendix shows, but there are no comparable sources on identities and lifestyles. Most insights into urbanism are the product of in-depth sociological studies of which there is a deficiency in developing countries. Such work as does exist needs to be evaluated critically since it is difficult if not impossible for researchers to shed their own cultural values and so study alien societies, and indeed their own, objectively and dispassionately. What it suggests, however, is that society in western Europe, much of North America and parts of Australasia, the

Middle East and South America is deeply permeated by urban values and dominated by urban institutions. Elsewhere the incidence of urbanism is limited or is non-existent. Vast tracts of Africa and Asia, and extensive areas in other parts of the developing world, are largely unaffected by urban influences. Rural ways of life predominate over large parts of the contemporary urban world.

Urbanism spreads through indigenous change and spatial diffusion. Urban lifestyles originate in cities in response to the opportunities and constraints afforded by place. From there they extend outwards to surrounding rural areas. Place influences and conditions behaviour because people in cities live in large numbers and at high densities in artificial environments, so they develop different patterns of association and living. They adopt lifestyles which are determined by location.

Urbanism can, however, be extended and exported to distant destinations well beyond the city via print, film, tape, disk, telephones, computers, radio and television. The introduction of satellite relays, long-distance telecommunication, audio and video cassettes and the Internet, has enabled the reach of these media to become global. They can be accessed over a large part of the Earth's surface by anyone who has access to the basic technology. Media material is rich in images, symbols and values which popularise and promote urban patterns of social and economic relationship. It establishes icons and reference points against which individuals and communities can define their identities. Advertising carries especially powerful meanings, as its purpose is to influence behaviour and promote sales. It seeks to define culture in terms of commodities such as cars, soft drinks, cigarettes, clothes and cosmetics and to encourage identity and belonging through consumption. An important feature of the mass media is that they portray and project lifestyles and modes of behaviour in cities which may differ radically from, and so clash fundamentally with, those that are the accepted norm in traditional societies. The world's rural areas are progressively being exposed to, and are being forced to come to terms with, urban influences both from within their own countries and from societies which may be both culturally, as well as geographically, distant.

The rapid spread of media-based urbanism over the last two decades has raised heated debates over impacts and implications. It has been viewed with alarm by some observers who worry that it is leading to a cultural homogenisation in which, ultimately, everyone everywhere will lead the same urban way of life. There is particular concern over qualitative aspects, with the spectre of urban images, attitudes and values flooding out of decadent western cities to the detriment of ancient and rich rural ways of life in distant developing countries. This perspective is associated with what some regard as the 'McDonaldization', 'Coca-Colaization' or 'Hollywoodization' of society (Cochrane, 1995: 250). It is readily conceptualised within the framework of 'cultural imperialism' or, even worse, 'American cultural imperialism'. It leads to fears of sameness and dumbing down, as exposure to global forces produces a uniform and debased urban culture.

The alternative point of view is that this model is too simple, as it underestimates the depth and resilience of traditional ways of life. Urban lifestyles, it is countered, are being extended across the globe within the framework of cultural pluralism: there is no one urban culture and so no single local response. Rural areas in developing countries are being exposed to outside influences and are resolving differences and conflicts between the global and the local in diverse and varied ways. They are generating hybrid forms rather than being sucked into a single cultural orbit. Examining the nature, spread and implications of urban culture, alongside the study of urban growth and urbanisation, is a key element in any attempt to understand the characteristics of the contemporary urban world.

Culture and community in an aspatial world

The interest in the future of communities and cultures in a world in which space was losing its importance arose out of the growing realisation during the 1960s that transport and telecommunications had far-reaching implications for the organisation of society at the global scale. It was at this time that the introduction of jet travel and a big increase in the size and quality of telephone services undermined the frictional effects of time and distance and effectively shrank the world to a manageable size. The urban consequences were first addressed by the communications guru, Marshall McLuhan, in his book *Understanding Media: The Extensions of Man* (1964). The key argument was that telecommunications were reintroducing the close interpersonal contacts that characterise rural living. Rather than an urban world, we shall all ultimately live in a 'global village' (Box 6.1).

Box 6.1 Marshall McLuhan and the 'global village'

The global village is seen as the creation of telecommunications technology. For Marshall McLuhan, history is made up of three stages. The first is the pre-literate or tribal stage in which people live close together and communicate orally. This is followed by the Gutenberg or individual stage in which communication takes place by the printed word and thinking is done in a linear-sequential pattern. The third is the neo-tribal or electric stage in which computers, television and other electronic communication media restore close inter-personal links. The significance of the latter is that it makes clustering of activities and people at particular places unnecessary:

> Before the huddle of the city there was the food gathering phase of man the hunter, even as men have now in the electric age returned psychically and socially to the nomad state. Now however it is called information gathering and data processing. But it is global and it ignores and replaces the form of the city that has therefore become obsolete. With instant electric technology the globe itself can never again be more than a village.
>
> (McLuhan, 1964: 366)

McLuhan's arguments echoed and extended those of Meier (1962), as set out in his communications theory of urban growth. Meier analysed the nature of the bond between two individuals and suggested that the quantum involved was the transaction: the exchange or transfer of information between sender and receiver. Bond formation between individuals, he argued, is facilitated by geographical proximity and by the acquisition and retention of knowledge, so that cities evolve primarily as a means of facilitating interpersonal communication. An important feature, indeed a major attraction of city life, is the time spent in public and professional, as opposed to private and family, life, so that shared symbols and experiences generate civic bonds which help to maintain and reinforce the cohesion of the city. Urban growth therefore takes place as cities develop a capacity to maintain and conserve information. The city is seen as an open communication system, resulting from, and held together by, a complex pattern of information exchanges.

Meier related urban growth not to changes in economic product or to the size of the social group, but to developments in communications technology. Social exchange in early settlements was primarily achieved by face-to-face contact and, with people attempting to maximise their chances of social interaction by locating close to the central zone of conflux, cities were both densely populated and compact. They were dominated by a small elite of priests and politicians who, through public meetings and assemblies, controlled the distribution of community information. With the progressive introduction of handwritten records, printing, publishing and broadcasting, communications technology complemented, and then largely replaced, face-to-face contact as the prime means of information dissemination. Modern cities are defined not simply in physical terms, but as social networks in space, created, maintained and manipulated by a wide range of communications media.

The most important characteristics of such networks is that they are defined by involvement and not by proximity. Although individuals live in a particular place and participate in community life in and around that place, it is interaction and not place that it the essence of city life. Individuals are increasingly able to maintain contacts with others on an interest basis and so be members of interest communities which are not territorially defined. Modern communication technologies open up a choice of lifestyles, once the prerogative of those who live in the city, to all. Propinquity is no longer a prerequisite for community.

For Webber (1964), the extent and range of participation in interest communities is a function of an individual's specialisation rather than place of residence. The more highly skilled that people are, or the more uncommon the information they hold, the more spatially dispersed are the members of their interest group and the greater the distances over which they interact with others. There is a hierarchical continuum in which the most highly specialised people are participants in interest communities that span the entire world; others, who are less specialised, seldom communicate with people outside the nation, but regularly interact with people in various parts of the country; others communicate almost exclusively with their neighbours (Figure 6.1). For example, research workers will be members of a wide range of interest communities and will spend more of their time participating in national and international scale communities than will the primary school teacher. Similarly, the upper limit of the interest community of school caretakers will in all probability be restricted to the city. All three will, however, be involved in more parochial interest groups in association with their roles as parents, or members of local clubs and societies.

For any given level of specialisation, Webber argued that there are a wide variety of interest communities whose members conduct their affairs within roughly the same spatial field or urban realm. Such urban

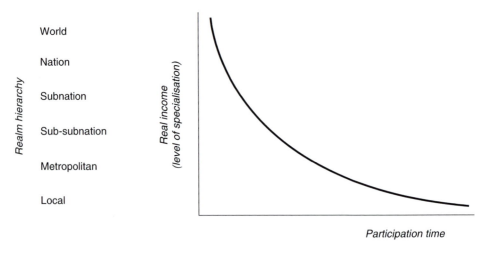

Figure 6.1 *Webber's concept of the structure of interest communities by realm*

realms are neither urban settlements nor territories, but heterogeneous groups of people communicating with each other through space. They are somewhat analogous to urban regions, but, contrasting with the vertical divisions of territory that are organised around cities and accord with the place conception of regions, urban space is divided horizontally into a hierarchy of non-place urban realms (Figure 6.2). Irrespective of location, people at different moments are participants in a number of different realms as they shift from one role to another, but only the most specialised people communicate across the entire nation and beyond. In this context the city is no longer a unitary place. Rather it is part of a whole array of shifting and interpenetrating realm spaces which exist at a variety of spatial scales from the local to the global.

The non-place global village as envisaged by McLuhan, Meier, Webber and others remains a remote prospect, despite the lapse of 40 years since its original conception. It is a captivating idea, eagerly embraced by technophiles, but pronounced variations across the world in the availability and use of telecommunications and information technology mean that it is a distant dream. Some people, most notably those in the highest levels in business, government and academia live communications-intensive lifestyles in a world in which, for them, time and space have ceased to have much meaning. They use air travel, telephones, faxes and computer networking to alternate regularly and repeatedly between interest communities at different scales up to the

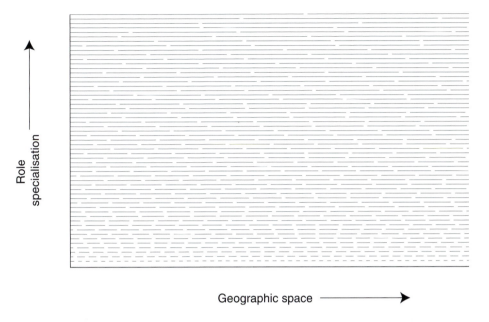

Role specialisation

Geographic space

Figure 6.2 *Webber's concept of role specialisation and its geographical relationships*

worldwide. For them, participation is everything; place is incidental. Such individuals approximate the archetypal citizen of the global village and may be pointers to the ways in which we shall all live our lives in the future. They are, however, very few in number even in the developed economies where role specialisation and technology are highly advanced.

Significant uptake of the technologies that might make the global village a reality is limited to a small number of the world's most affluent nations. The majority of the population in these areas has a mobile phone and uses the Internet. Elsewhere, the information age has yet to dawn, as penetration levels for information technologies are negligible. Some 16 per cent of people are believed to have owned mobile phones in 2001 (Table 6.1). The level was above 75 per cent in Taiwan, Hong Kong, Israel and several countries in north-western Europe, but it was only 3 per cent in Africa. The pattern of Internet use is the same, although the variations are more pronounced. Overall, some 8 per cent of people were Internet users in 2001. The level was above 50 per cent in North America and parts of north-western Europe and eastern Asia. Internet use in Africa was below 1 per cent. The pattern of use of personal computers was very similar.

Table 6.1 *Information technologies: access and use, 2001*

Cellular mobile phone users (%)	Personal computer users (%)	Internet users (%)
Taiwan (97)	USA (62)	Iceland (67)
Luxembourg (97)	Sweden (56)	Singapore (61)
Hong Kong (85)	Australia (52)	Norway (59)
Italy (84)	Luxembourg (51)	Sweden (51)
Austria (81)	Switzerland (50)	Korea (51)
Israel (81)	Denmark (43)	USA (50)
Sweden (79)	Netherlands (42)	Hong Kong (45)
UK (78)	Iceland (41)	Japan (45)
Africa (3)	*Africa* (1)	*Africa* (1)
Asia (9)	*Asia* (3)	*Asia* (4)
World (16)	*World* (8)	*World* (8)

Source: www.itu.inter

Passive citizenship of the global village, however, is both substantial, widespread and increasing rapidly. An argument can be made that a global urban society is being created in the form of a worldwide community of radio listeners, and television and video viewers who see and interact with the city as an image, not a place. Their urban experiences are predominantly electronic. Most broadcasting takes place from, and is about, cities, so this armchair audience receives a diet of programmes and material which projects and in many cases endorses and glamorises urban ways of living. Viewers are exposed to pictures of cities as places, and to the activities and dialogue about the lifestyles and patterns of behaviour which take place within them. Urban fashions and fads are publicised and the relationships which underpin urban communities are explored and popularised. Fictional characters from *Neighbours*, *Dallas* and *Dynasty* become urban role models, and action, violence and extremism, which are visually exciting, are emphasised. The global impinges ever more on the local as viewing times and audiences increase. Traditional values are in retreat as mass media urbanism progressively permeates the most distant and remote rural societies.

The extent to which such imagery is extending urbanism across the globe and, if it is, what form that urbanism takes is, however, a matter for debate and dispute. Global urbanism is a simple, superficially plausible, but highly questionable concept. Three key questions are involved. The first concerns the size and character of the global market and surrounds the extent to which media products are available to and are consumed by worldwide audiences. The second focuses upon the urban content and significance of this material. The third is about the impact of this content on the behaviour and ways of life of listeners and viewers.

Such questions cannot, however, be answered definitively, due to the limited research on global mass media and their impacts. Two crucial deficiencies are work on the flow of communication between countries and on the cultural biases in those flows. Both have been written about extensively, but there have been few empirical studies, and by far the great bulk of that literature consists of polemical essays unenlightened by facts. Against this background the following sections attempt to synthesise the largely speculative views that have been advanced concerning the consequences for urbanism of the growth of global culture.

Global media services

Media urbanism is essentially a creation of modern telecommunications, although it has its roots in the film, telephony and television technologies of the early twentieth century. It traces its origins to the introduction of the first commercial radio broadcasts in Pittsburgh in 1926 and the first public television service by the BBC in London in 1936. Together these developments initiated a process through which the world has become progressively interlinked for instantaneous communication (Box 6.2). Today, commercial radio and television services are ubiquitous. The very few countries which do not have their own television service are in Africa and include Botswana, Cameroon, Central African Republic, Chad, Gambia, Malawi, Rwanda and Western Sahara. Several of these are very small and share the television service of a larger neighbour.

The size of television audiences around the world, however, varies widely because of the high costs of installation both of broadcasting stations and receivers in the home. Most of the television receivers

Box 6.2 Global television

Global television was made possible by the launch in 1965 of INTELSAT (Early Bird), the first commercial satellite. Early Bird doubled the telephone capacity between the USA and Europe and for the first time enabled live transmissions between the two continents. Today there are over 200 communications satellites in orbit and global television is a reality. The world's six largest media organisations – Rupert Murdoch's News International, Time Warner, Disney, Bertellsman, TCI and Viacom – all have a global presence, broadcasting around the world via satellite subsidiaries such as BSkyB, RTL, Fox and CNN. The number of sets rose from 192 million in 1965 to 1.4 billion in 1997 (UNESCO, 1999). Some 500 million people watched the moon landing in 1969 and around a billion saw some part of the 1976 Olympics in Montreal, while over four billion probably watched the 2000 Games in Sydney.

Television only began to reach a mass market in the last 30 years, although the technology was invented over 70 years ago. Time lags involved in establishing networks of transmitting stations, in developing programming schedules and in the increase in the ownership of receivers meant that the spread of television was slow. Television became a mass medium in the USA and the UK in the mid 1950s, but these are exceptions. In 1954 there were only 100,000 sets in France and fewer than 90,000 in West Germany, a density of ownership that was fewer than three sets per 1,000 people. Early experiments with television in China did not begin until 1956 and it was not until 1978 that the China Central Television organisation, which coordinates broadcasting over most of the country, was established. Television in China did not start to enter the homes of working people until the 1980s (Lull and Se-Wen, 1988). Venezuela is similar in that television did not become a national service until the late 1960s (Barrios, 1988).

presently in use are in the developed world and around one-quarter of the total is in the USA. Among the ten countries with the largest number of receivers, only China, the world's most populous country, and Brazil are in the developing world. There are more television receivers in use in the UK than in the whole of Africa, and more in Japan than in South America (UNESCO, 1999). The extent to which television is presently a communication medium of the developed world is underlined forcefully when allowance is made for population (Figure 6.3). The number of televisions per 1,000 inhabitants is high in North America, Japan, Australia and parts of Europe, the Middle East and South America, but is very low throughout most of Africa and Asia. Among the major countries for which data are available, there are fewer than five televisions per 1,000 inhabitants in the Central African Republic, Chad,

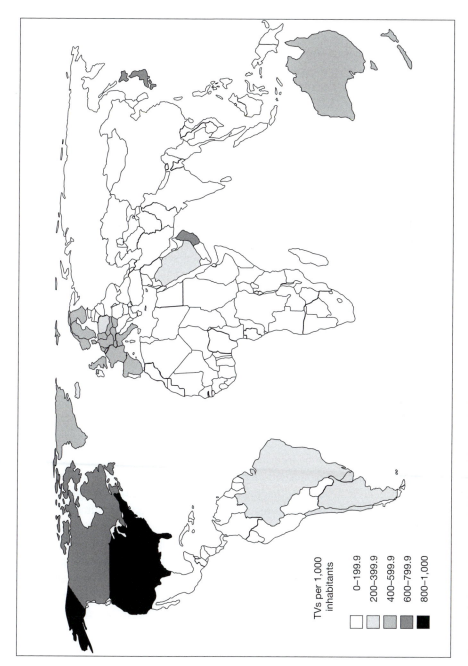

TVs per 1,000
inhabitants

  0–199.9

  200–399.9

  400–599.9

  600–799.9

  800–1,000

Figure 6.3 Televisions per 1,000 inhabitants, 2001

Ethiopia, Mali, Mozambique, Tanzania, Zaire and Myanmar. In these and other developing countries, the lack of electricity services in rural areas is a major impediment to the spread of television.

It is important to emphasise, however, that a simple mapping of the number of sets per head of population provides only a very crude measure of the local importance of television. Figure 6.3 overstates the position in the USA and Japan since, in houses in which there is a set in each room, most are used infrequently. Conversely, in countries in the developing world, television is watched by many more people than the number of receivers suggests. Multiplier effects are high because even a small set can be viewed by several people at a time. With a newspaper, even two people cannot read a single copy comfortably. The importance of the extended family in traditional societies makes a home with a television a place where relatives and neighbours gather to watch and so serves to extend any impacts. In the major towns almost every social or sports club has a colour television set. Television enjoys considerable novelty appeal in many parts of the world and is the principal means of entertaining and informing in societies in which the majority of the population is illiterate. It is more influential than radio because the information content of pictures is far greater than that of sound. What is transmitted, good or bad, home produced or foreign, reaches billions of viewers irrespective of age, colour, sex, religion or level of literacy.

Urban imagery and values

Television tends to be rich in urban content and imagery because most broadcasting companies are city-based and so direct their attention towards civic affairs and happenings. Cities are places in which the major political events take place, so they are the focus of attention in most news and current affairs broadcasting. The extremes of urban poverty and affluence are a common focus of inquiry in documentary programming, while the cultural importance of cities is projected through the broadcasting of theatre and opera, which are essentially urban art forms. The concentrated activity and action which takes place in cities is visually exciting and makes for supposedly 'good' television drama. Urban settings form the backdrop to most soaps and *telenovelas* and the values of competitiveness, conspicuous consumption, selfishness and infidelity, which are commonly portrayed in series such as *Dallas* and *Dynasty*, are readily taken to be urban values.

Irrespective of where they live, audiences around the world are fed a broadly similar diet of television. The same types of programmes are scheduled at the same times of the day. Broadcasting tends to follow the same basic pattern because it is geared to people's daily lives. Soap operas and quiz shows account for most of the daytime slots, while children's programmes predominate in the early evening. These are followed by family viewing, the mid-evening news, drama, sport and adult television. The significance of this standard format is that it generates demands for particular types of programming, much of which is international in origin.

Although most countries have their own television services, few are able to generate sufficient output to fill their schedules, so they rely heavily on imports to make up the difference. It is estimated that around one-third of total television programme time across the world is imported material. The USA is the biggest importer of television programmes, but it is the smallest importer in relation to its total television programme output. Many countries rely on imports for more than half of their programming and it seems likely that most of the African countries fall into this category (Box 6.3).

The demand for imports is high because of the economics of contemporary broadcasting. Television is a western invention which has spread slowly and unevenly throughout the world over the last half-century. Much of the equipment, most of the know-how and many of the programmes still originate from Europe and North America. This dominance may be seen as a form of neo-imperialism, but the reasons

Box 6.3 Television in Venezuela

An example of a country that makes significant use of viewing imports is Venezuela, where it is estimated that some 51 per cent of television is foreign material, the majority from the USA. Television programming in Venezuela has some similarities in its composition to the major American networks in that it is possible each week to watch major serials, including *Falcon Crest, Dynasty, Dallas, Miami Vice, Kojak* and *He-Man*. Nearly half of daily programming is made up of American films and children's programmes. Many of the tele-images that Venezuelans receive at home are from Los Angeles' streets, Miami's beaches, Dallas's offices, Manhattan's avenues and the suburban houses of American middle-class families.

are financial rather than overtly political. Low-income countries with embryonic television services have little choice but to buy programmes from abroad. Few have the resources to produce even five hours of television a day. Even if subtitles or dubbing are required, the low price of many imported programmes and the lack of local suppliers mean that station managers have to rely heavily on imports (Box 6.4).

Global urban culture or cultures?

The implications of the mass media for the extension of urban cultures are much disputed. A large and rapidly expanding global audience is fed a diet of television rich in imported products, but it is far from clear how this influences identities, affiliations and behaviours. The key arguments surround the power and effectiveness of television as an agency of social transformation and the directions that such media-induced changes are taking. Two alternative perspectives can be identified in the recent literature. The more traditional sees television as imposing a uniform urbanism of a western variety upon, and to the

Box 6.4 English language broadcasting

Producers from the USA, the UK, Canada and Australasia have a competitive advantage in the global television market because they work predominantly in the English language. English is the main international language, not because it is the first language of most people, but because it is their second, third or fourth language. About one-fifth of the world's six billion people are believed to use English on a daily basis. This still puts English behind Mandarin Chinese in numbers, but Chinese is not an international language. It is largely restricted to the mainland of China, where it is the language mainly of native speakers. Other languages such as Spanish, Hindi, Arabic and Portuguese may soon surpass English in terms of numbers of native speakers, but none so far have the international status of English. It is the single official language in 25 countries and a co-official language in 17 more. English is institutionalised as the language of government and education in many countries in which there is no common indigenous language. It is the primary linking language in several countries such as India, where there are many local or regional languages. Although only 3 per cent of the population is actually bilingual in English and Indian languages, much of government and higher education, and approximately half of all the books published in India, are in English. Global pre-eminence is reinforced because English is the main language of science and technology. It is the language for which most computer software is written.

detriment of, indigenous cultures. The more recent highlights the complexity of television flows and, as countries increasingly draw their programmes from a wide variety of sources, argues that the effects will be many and varied.

The traditional view is that mass media are creating a homogeneous urban culture which is being spread to all parts of the globe by television and related technologies. This culture is heavily infused with western urban imagery which, as it washes across the world, is progressively displacing residual ruralism (Schiller, 1976; Mattelart, 1979). As such it can be conceptualised as 'cultural imperialism' in the sense that media dominance is seen as the logical corollary of economic and political hegemony.

This model rests upon two assumptions. The first is that television has the power to override all the other traditions and institutions that shape local society. The second is the supposed ubiquity and popularity of western, and especially American, television and the belief that it will displace local media forms to the detriment of parochial cultures. Despite the highly questionable bases for these assumptions, 'it is a thesis which has been spoken about in countless conferences, seminars, books and pamphlets and mouthed so loudly that it has been transferred from postulate to certain truth' (Tracey, 1993: 164). It has alarmed politicians, such as the French Minister of Culture who, in 1982, identified *Dallas* as a national threat and called for a crusade against financial and intellectual imperialism that 'no longer grabs territory, or rarely, but grabs consciousness, ways of thinking, ways of living' (ibid.: 178). Machimura (1998) notes a similar response by local government in Tokyo, which was concerned about the detrimental effects of the mass media and attempted to stress the merits of local culture so as to engineer the 're-Japanisation' or 're-Asianisation' of the city. The long-term implication is that the world will be subsumed by a homogeneous urbanism – a polyglot urban culture in which everyone everywhere will live a way of life dominated by urban values.

Recent research, however, suggests that mass media are in fact contributing to the creation of highly varied patterns of urbanism at the global scale, and that the ideas of cultural imperialism lack empirical support (Sreberny-Mohammadi, 1991; McEwan, 2001). If culture is a system of shared meanings, then there are many systems of shared meanings. The concept of a plural urban culture is grounded in analyses of the international market in media materials which highlight the complex pattern of production, content and distribution. The USA and other western nations are no longer the only producers of media materials. Moreover, their products do not always have a dominant presence in the countries into which they are imported, nor do they necessarily attract large audiences. Where there is a choice, domestic programmes are invariably preferred. Local output is not threatened by imports and, in most countries, is rising rapidly. It may project images of life in the city at the expense of those in the countryside, but such urbanism is extended predominantly within the context of national rather than global culture.

There is growing evidence to support the plural urban culture model. The most basic is the sheer increase in the volume of media material being generated in the developing world. Mexico and Brazil are major producers and distributors of television programmes, with TV Globo, the Brazilian network, exporting *telenovelas* to 128 countries, including Cuba, China and the former USSR. The Indian film industry is the most productive in the world, making nearly 900 films in 1998 (UNESCO, 1999). Most are Hindu epics in which traditional cultural themes and values are explored and reaffirmed. Egypt is an emerging centre of film-making in the Arab world.

Increasingly complex flows and exchanges of television programmes are revealed in studies of the global television market that were undertaken for UNESCO by Ramsdale (UNESCO, 1999). These challenge the established view that western urban values are spreading across the world via a strong one-way flow of cultural goods from the north Atlantic basin to developing countries. For example, in Arab countries, 42 per cent of television is imported, of which one-third comes from other Arab states. The United Arab Emirates, Egypt, Saudi Arabia and Kuwait are the main suppliers. France provides about 13 per cent and the UK, Japan and Germany between 5 and 7 per cent each. Similarly, in Africa, about 40 per cent of programmes are imported, although again there are wide differences in volume between individual countries. Some 50 per cent come from the USA, 25 per cent from Germany and the rest principally from western Europe. In South Africa, 30 per cent of programmes are imported: 54 per cent from the USA, 30 per cent from the United Kingdom, 9 per cent from France, 5 per cent from Austria and 3 per cent from Canada. In Europe, more than 40 per cent of imported programmes originate within other countries in the region itself.

These, and similar data from other parts of the world, point to the existence of highly intricate patterns of media production and exchange. The number of producers is increasing and regional markets operate and are growing strongly. Close links exist between suppliers and consumers within particular language groups, so elements of cultural distinctiveness are maintained. Portugal, for example, is a major importer of Brazilian television programmes (Sinclair *et al.*, 1996). Flows of videos are broadly similar to those of television (Alvarado, 1989). The USA may have dominated the market when film and television were in their infancy, but the contemporary situation is one of multilateral production and trade in media products.

A second argument is that imported material is rarely as popular as domestic programmes and is frequently used to fill the off-peak slots (Box 6.5). Thus, in Japan, *Dallas* was introduced in October 1981 and went to only a 3 per cent market share in December. It never remotely compared with the popularity of *Oshin*, a locally produced six-days-a-week fifteen-minute serial drama which regularly attracted over half of the national audience. Locally produced '*telenovelas*' are far more popular in South America than imported American soaps. For example, in the mid 1980s, TV Globo broadcast a *telenovela* called *Roques Santeriro* which regularly captured 90 per cent of the national market. Similarly in Venezuela, while US and Brazilian soaps were popular they did not compete for the prime-time spots filled by local productions such as *Cristal* and *Los Donatti* (Patterson, 1987).

Box 6.5 Viewing preferences in the Netherlands

The preference for domestic productions is strong even in the Netherlands, where there are many competing channels and a sophisticated and highly educated audience. Dutch viewers can receive television from several neighbouring countries, including the UK, and as many of them can understand French, German and English, there is a wide range of choice. The work of Bekkers (1987), however, shows that, even in this extreme case of a pan-European market, the preference for domestic channels is strong. The two leading Netherlands channels have an overwhelming market share that increased during the 1980s in spite of competition from satellite broadcasting.

The popularity of domestic television emphasises the continuing strength of national cultures and the power of the languages and traditions through which they are perpetuated.

A third argument is that global messages are evaluated and are absorbed selectively in ways relevant to local experiences through a process which has been described as 'indigenization' (Appadurai, 1996). Evidence exists that people choose meanings and symbols from global media that they see as important and use them to enrich aspects of their traditional cultures and ways of life. An example is the response of locals when television was brought to a town in Nigeria, as reported by Hannerz (1997). Many elements in the broadcasts were deemed by locals to be of little interest, but some were responded to positively and clearly affected attitudes and behaviour. The consequence was a cultural transformation in the form of a selective fusion, through what Hannerz termed a process of 'creolisation'.

The evidence is far from equivocal, but it identifies a different urban present and points to a different urban future to that suggested by cultural imperialists. The view of Lull (1995) is that, despite the worldwide reach of technology, we do not and will not live in a global village where an all-encompassing and uniform western urbanism replaces outdated and unwanted rural ways of life. Mass media are extending the reach of urban values, but they are not impacting upon local cultures in the same way. The interaction between global and local results in a range of hybrid cultural forms. Just as living in a city does not create a homogeneous urban lifestyle in a single society, so the worldwide transmission of information produces richness and variety of urban responses at the global scale. There is no single urban culture; there are many.

Conclusion

The growth and spread of urbanism amounts to social change on a vast scale. It originated in the industrial metropolis, where the concentration of people from diverse backgrounds gave rise to and enforced a variety of lifestyles that were fundamentally different to those which existed in rural areas. Such patterns of association and behaviour were initially seen as the products of social breakdown, as traditional forms of social control, based upon the membership of primary groups, were destroyed by the size, density and heterogeneity of urban populations. Later research pointed out that urbanism was not restricted to cities, but could be spread across the globe in the form of images and values carried by the mass media. Large parts of the world are presently permeated by urban values and urbanism touches and impinges upon traditional value systems in many of the most remote and traditional rural localities.

Worldwide social relations have intensified, so that distant areas are linked in such a way that local happenings are shaped by events occurring many miles away, and vice versa (Giddens, 1990). The extension of urban culture is likely to increase as the numbers and proportion of the world's population living in urban places, and the power of the media to spread urban values, rise.

The global spread of urban values was traditionally viewed by many in largely negative terms. At best it was equated with undesirable modernism. At worst it was seen within the framework of 'cultural imperialism', alongside the rise of the transnational corporation, as a means of maintaining and extending the pre-eminence of western capitalism. A central assumption was that western, and especially American, urban values such as competitiveness, consumerism and individualism, which were expressed explicitly in a variety of media genre as well as indirectly through advertising, were being exported to, and so were devaluing, the rich and ancient cultures of many developing world countries. Fears of 'cultural homogenisation' were voiced and arguments were made for 'cultural dissociation' from the global capitalist system as a means of ensuring autonomous development and the protection of indigenous cultures (Sreberny-Mohammadi, 1991).

Today, it is possible to see the spread of urban values across the globe more positively. The international media market is far larger and more complex than that suggested by the cultural imperialists. Western media domination has given way to multiple suppliers and complex cross-flows of media products. More developing world nations are producing and exporting media materials, so the variety of urban imagery in the world marketplace is increased. Rural values and ways of life are under attack, but from a variety of directions, both domestic and foreign, and in a number of complex ways. The picture is one of increasing urbanism, but not necessarily greater social uniformity. Rather than a single global village with a homogeneous culture, the trend is towards a plural worldwide urban society.

Recommended reading

Crang, M. (1998) *Cultural Geography*, London: Routledge. An excellent discussion of the meaning of culture and the ways in which geographers think about culture. The book explores local and global cultures of production and consumption, and ideas about place, identity and nationhood.

Curran, J. and Gurevitch, M. (1992) *Mass Media and Society*, London: Arnold. A collection of review papers on media production and impacts. It includes an especially useful contribution from Sreberny-Mohammadi on global–local relationships in international communication.

Daniels, P., Bradshaw, M., Shaw, D. and Sidaway, J. (eds) (2001) *Human Geography: Issues for the Twenty-first Century*, Harlow: Pearson. The excellent chapter by McEwan on 'Geography, Culture and Global Change' covers debates about the nature of global culture and its local impacts in depth and in detail.

Holloway, L. and Hubbard, P. (2001) *People and Place: The Extraordinary Geographies of Everyday Life*, Harlow: Prentice Hall. This book explores in a stimulating and informative manner the ways in which people think about, represent and relate to the places in which they live and work. Of particular relevance is the emphasis which is placed upon the importance of culture in shaping people's lifestyles, behaviour and identities.

Massey, D. and Jess, P. (eds) (1995) *A Place in the World? Places, Cultures and Globalisation*, Oxford: Oxford University Press. A comprehensive Open University introductory text which examines the conceptualisation of place and how this relates to identity and cultures.

Tomlinson, J. (1999) *Globalisation and Culture*, Cambridge: Polity. An advanced analysis of the relationship between globalisation processes and contemporary cultural change. The book is for those who wish to consider cultural issues from the standpoint of social theory.

Webber, M. M., Dyckham, J. W., Foley, D. L., Guttenburg, A. Z., Wheaton, W. L. C. and Wurster, C. B. (1964) *Explorations into Urban Structure*, Philadelphia: University of Pennsylvania Press. An early, but still relevant, theoretical consideration of the relationship between interaction and spatial structure. Webber's chapter on the urban place and the non-place urban realm is especially stimulating.

Zukin, S. (1995) *The Culture of Cities*, Oxford: Blackwell. An in-depth examination of the nature and meaning of urban cultures based upon American experiences and examples. Particular attention is paid to the 'symbolic economy' as expressed by life on the streets of New York.

Key web sites

www.itu.int   This is the site for the International Telecommunication Union. It provides access to a wide range of information on the availability and use of telecommunications and information technologies.

www.unesco.org   This is the site for the United Nations Educational, Scientific and Cultural Organisation. It provides access to the biennial *World Information and Communications Report*.

Topics for discussion

1 'We are all citizens of a global urban society'. Discuss.

2 Examine critically the value of the concept of the global village in understanding lifestyles in the contemporary world.

3 'It is interaction and not place that is the essence of city life' (Webber,
1964). Discuss.

4 Assess the impact of telecommunications and information technologies upon
lifestyles in cities.

5 Assess, with reference to last night's programmes, the extent to which
television projects and popularises urban ways of living.

7 **World cities**

By the end of this chapter you should:

- understand the concept of world cities and the debates which surround the meaning and significance of the term;
- be familiar with the functions of world cities and the roles they play in the world economy;
- be aware of the issues which surround the specification and measurement of world city status;
- understand the reasons for the rise of world cities.

Introduction

The urban world is dominated by a small number of centres that are the command and control points for global capitalism, the world's dominant economic system. Such centres are distinguished not by their size or their status as capital cities of large countries, but by the range and extent of their economic power. They are the locations for the key individuals, institutions and organisations that manage, manipulate, dictate and determine the formation and reproduction of capitalism across the world. These attributes give such cities a disproportionate and exceptional importance, so that they occupy dominant positions in the global urban hierarchy. So pre-eminent is their status and so powerful and pervasive are their influences that many analysts argue that they merit designation as world cities.

Global economic importance and connectivity are the key criteria for world city status. Most observers today, taking their lead from the ideas of Friedmann (1986), see world cities as the decision-making and control points for the world economy (Beaverstock *et al.*, 1999). World cities are distinguished by their roles as sites for the accumulation and concentration of capital and as places from which its distribution and circulation are organised and managed. They are favoured locations for the institutions of international production and consumption and the individuals and agencies that support and facilitate these activities. Function rather than size is critical. World cities are places in and from which global business, finance, trade and government are orchestrated and arranged.

The command and control role is reflected in the activities that are typically located in world cities. These include corporate management, banking, finance, legal services, accounting, technical consulting, telecommunications, computing, international transportation, research and higher education. Concentrations of such functions are common in capital cities that serve as the highest-order service centres for their national urban systems. They are especially pronounced in the principal business centres of the richest and most advanced economies. The number and range of services do not of themselves, however, denote world status since outlook and orientation are key considerations. What distinguishes world cities is that they provide services for a world market and not merely for domestic or regional consumption.

World cities are characterised by concentrations of headquarters of global corporations, commodity, currency and securities exchanges, and head offices of producer services organisations. They are major centres for international government and administration and are principal junctions on the global conference and convention circuit. The importance and distinctiveness of these activities are conveyed by office and convention buildings of distinctive architectural design (Zukin, 1992). They give world cities an identity that symbolises their economic dominance (Plate 7.1). The presence of global functions and institutions means that world cities have more in common with each other than they have with urban centres in their own countries and with places of similar size elsewhere. The strength of their interlinkage means that it is often as easy to travel among them as it is to reach secondary centres in the same country or in adjacent territories. World cities are the principal foci for global business travel and telecommunications networks.

The presence of large numbers of members of the transnational corporate and producer service class means, it is argued, that world cities have a distinctive sociology that is expressed in terms of occupation, income and ways of life of their residents. They are places of social polarisation, which may be conceptualised as 'dual cities' (Castells, 1989) or 'divided cities' (Fainstein et al., 1992). Those who work in the organisations and institutions that sustain world city functions constitute a well-educated, socially mobile, footloose and highly paid elite. They are cosmopolitan in origin and global in outlook. Their corporate, diplomatic and professional skills are well developed,

Plate 7.1 *The skyline of a global city: the buildings of Manhattan house the head offices of global corporations, financial institutions and advanced producer service organisations*

highly prized and generously rewarded. The presence of this group of international service sector workers stretches the social profile more than that in other cities, and creates a wide gap with those at the opposite end of the spectrum. A low-skilled and low-paid working class that services the international service sector exists alongside, but is well separated from the community of global professionals. It typically includes large numbers from ethnic minority backgrounds. World cities are places of exceptional wealth and affluence, but they are also places of severe disadvantage and deprivation.

The degree and significance of social polarisation are, however, disputed, especially by those who are most familiar with urban places outside the USA. One argument is that social inequalities are no greater in world cities than in other major cities and in any case are probably no more pronounced today than they were in the past (Abu-Lughod, 1995). Detailed empirical research is required to measure and explain spatial and temporal variations in social opportunity. A second argument is that the degree of social polarisation is overstated because there is too strong a focus in the research literature upon American cities. The low-wage service economy is smaller in the major cities of Europe because of better worker protection and welfare provision (Hamnett, 1994, 1996). A third area of debate surrounds attitudes to social conditions. World cities generate enormous opportunities which many people grasp, not least members of the growing army of international migrants (Findlay et al., 1996). The fact that some do not, it can be argued, is as much a reflection of personal failure as it is of defective distributional mechanisms within the economies of world cities.

The hierarchy of world cities

The places that are recognised as having world city status vary according to the mix of diagnostic measures used. London, New York, Tokyo and Paris are accorded this accolade in the majority of the 29 classifications that were published in the research literature between 1972 and 1998 (Beaverstock et al., 1999). The earliest of these studies is that of Hall (1966). In his book, *The World Cities*, he identified a set of places in which 'a quite disproportionate part of the world's most important business is conducted' (1966: 7). Hall distinguished world cities from other places of great population and wealth because they were major centres of political power, seats of national and international government and concentrations of related professional, trade union, employers' federation and corporate concerns. They were also centres of trade, finance and communication. Characteristically they were great ports, distributing imported goods to all parts of their countries and in return receiving commodities for export to the other nations of the world. Within each country they were the focus of road and rail networks and the sites of major international airports. Such activities gave the cities distinctive social and economic characteristics that were reflected in their status as centres of professional talent in the fields of medicine, higher education, research, culture and the arts. They were known for their universities, hospitals, concert halls and museums. On these bases, the places recognised by Hall as being world cities in 1966 were London, Paris, Randstad, Rhine-Ruhr, Moscow, New York and Tokyo.

A hierarchy of world city-ness today, consisting of three major tiers, is identified by the University of Loughborough's Globalisation and World City Research Group (Table 7.1). It reflects the contemporary view of Sassen (1991: 126) that world cities are 'postindustrial production sites' in which global corporate services and financial services are developed and provided. The classification is based upon a scoring of places on a 12-point scale according to their importance as providers of global advertising, banking, legal services and accountancy (Beaverstock *et al.*, 1999). At the apex are ten full-service world cities, each of global significance in all four service areas. London, New York, Paris and Tokyo score the highest, but Chicago, Frankfurt, Hong Kong, Los Angeles, Milan and Singapore are also included in this stratum. A second tier of ten major world cities, headed by San Francisco, Sydney, Toronto and Zurich, is of global significance in three of the four key world city functions. Beneath these in the hierarchy are 35 minor world cities, each of global importance in two of the key service functions. A further 55 places are identified which have some evidence of world city formation, but which do not merit a higher classification.

World cities have particular product and service specialisms, which reflect their history, location and the size and character of their national economies. In this respect they are complementary rather than in competition. New York, reflecting the manufacturing strength of the USA economy, is the principal locus of global corporate power. It is the main centre of global political and military power and a place from

Table 7.1 *The world city hierarchy*

Score	Cities
Full-service world cities	
12	London, New York, Paris, Tokyo
10	Chicago, Frankfurt, Hong Kong, Los Angeles, Milan, Singapore
Major world cities	
9	San Francisco, Sydney, Toronto, Zurich
8	Brussels, Madrid, Mexico City, São Paulo
7	Moscow, Seoul
Minor world cities	
6	Amsterdam, Boston, Caracas, Dallas, Düsseldorf, Geneva, Houston, Jakarta, Johannesburg, Melbourne, Osaka, Prague, Santiago, Taipei, Washington
5	Bangkok, Beijing, Montreal, Rome, Stockholm, Warsaw
4	Atlanta, Barcelona, Berlin, Budapest, Buenos Aires, Copenhagen, Hamburg, Istanbul, Kuala Lumpur, Manila, Miami, Minneapolis, Munich, Shanghai

Source: Beaverstock *et al.* (1999).

which a sizeable component of global production and consumption is controlled. Tokyo is a world city principally because of the post-war success of the Japanese economy. Its status is largely self-generated because it is physically remote from the western countries that are the traditional centres of the world economy, and because it was closed to the influx of foreign investment and immigration for many years. As the main export point for national financial wealth, it has a strong orientation towards its domestic sources of money supply as well as to world financial markets (Machimura, 1992). The manufacturing economy of the UK is comparatively small, but London is a world city because it is the principal supplier of financial and producer services to global markets, a role that it developed as the hub of the British Empire (King, 1989, 1990). It is also the cultural centre of the English-speaking world. Though less important as a corporate and financial centre, Paris is the most popular location for the headquarters of international organisations and for international conventions, a position that owes much to its many architectural splendours and gastronomic attractions (Knight and Gappert, 1989).

Although the concentration of capitalist functions in a small number of cities is not in dispute, many observers question the extent to which a handful of cities can and do perform a worldwide role. The concept of world cities is attractive, but the empirical substance is contested. One argument, as discussed in Chapter 2, is that the world urban system itself may be a misnomer, as so many countries and areas lie beyond its reach. For example, urban systems in large parts of Africa are presently embryonic and are poorly integrated. There are vast rural territories housing millions of people, many sustained by subsistence agriculture, who exist beyond or at best on the extreme margins of the domestic, let alone the global, urban economy. Their world is highly localised and their 'world' cities are nearby market towns. It is difficult to see how the lives of these people are affected, even indirectly, by the business that is transacted in New York, London, Tokyo or Paris.

A second argument is that it is easy to overstate the scale and extent of the power that is discharged from world cities. They may have a concentration of high-level decision-making functions, but major questions surround the extent to which they influence, let alone command and control, the global economy. One reason for serious reservation is that capitalist institutions in world cities are themselves shaped by global politics and economics. They seem to be all-powerful, but there are world events, such as the terrorist attacks on the USA on 11 September 2001 and the conflict with Iraq in 2003, which they were unable to prevent and which have a significant effect upon them. There is more to this debate than the validity and meaning of the 'world city' label and which places are and are not included. It addresses the key issues of how far the contemporary world hangs together as an economic and an urban system, and how far that system is dominated by and orchestrated from a small number of points of capitalist leverage.

Proponents of world cities argue that their rise both reflects and has made possible the emergence of the world economy. Three factors are held to be responsible: the growth of the number and range of the institutions of global capital, their geographical concentration, and the recent extension of global reach via telecommunications and transport. It is in relation to these factors that the emergence and role of so-called world cities, exemplified by London, New York, Tokyo and Paris, can best be analysed and understood.

The growth and activities of the institutions of global capital

World cities exist because they are the chosen locations for the agencies of global capitalism. They are places in which the principal business functions have developed and are concentrated and from which global corporate, financial and political control is exercised. A wide range of organisations is responsible for the accumulation and reproduction of global capitalism, including those concerned with goods production, financing and the provision of advanced producer and personal services (Table 7.2). It is the combined presence of such activities and the global roles that they perform that distinguishes world cities from mere national or regional centres.

Transnational corporations are the most important agencies and their proliferation and increase in size are both causes and consequences of the emergence of the world economy, as Chapter 4 has shown. The task of administering and managing global production requires elaborate corporate bureaucracies that are typically organised in a hierarchy of offices. At the apex is the global head office, normally in the country of origin, where strategic decision-making takes place and from where corporate empires are coordinated and controlled. It is here that the long-term future of the corporation is shaped and where operating structures and production targets for the constituent parts of the corporation are determined. Implementation of such corporate policies is typically the responsibility of regional head offices in each country or territory in which there is a significant production or marketing presence. An important role for staff at regional head offices is to integrate the corporation into the local business culture so as to minimise the potential sensitivities that surround external ownership and the remission of profits. Tactical decisions concerning day-to-day operations are typically taken at plant level by managers whose responsibility is to organise and maintain high levels of profitable production. Their job involves managing the receipt of raw materials and component parts, organising production runs and arranging for the dispatch of finished goods to the market (Box 7.1).

The growth in international production has been made possible by the development of an elaborate system of global finance that is organised through a number of different markets, and is mediated and controlled by a wide variety of finance capital institutions (Table 7.2). Complexity is the overriding characteristic and the distinctions between the various types of financial commodities and the markets and sub-markets in

Table 7.2 *Activities and organisations of global capitalism*

Commodity	Activities	Market structure	Organisations
Manufactured goods	Industrial production	World markets	Transnational corporations
Money	Borrowing, lending	Wholesale money markets, foreign exchange markets, Eurocurrency markets	Banks, discount houses, foreign exchange dealers
Financial securities	Securing of debt, speculation	Primary and secondary bond markets, financial futures markets	Banks and securities houses
Currency	Change, speculation, risk avoidance	Foreign exchange markets, Eurocurrency markets	Banks, foreign exchange dealers
Stocks and shares	Issuing, broking	Stock exchanges	Issuing houses, stockbrokers
Raw materials	Merchanting, broking	Commodity markets, futures markets	Brokers, merchant banks
Insurance	Underwriting	Insurance markets	Lloyd's and other insurance organisations
Freight	Chartering	Shipping exchanges	Chartering companies
Accountancy, legal, tax, advertising, public relations, management consultancy services, etc.	Provision of profesional services	World markets	Companies
Transport	Ticketing, carriage	World markets	Travel agencies, airlines, rail companies, car rental companies
Hotel	Provision of food and accommodation	World markets	Hotels and restaurants
Personal finance	Provision of cash, currency and credit	World markets	Banks, credit and charge card companies

Box 7.1 AGCO

The organisational and geographical structure of transnational corporations is illustrated by the example of AGCO Corporation, one of the world's largest designers, manufacturers and distributors of agricultural equipment. The company headquarters are in Duluth, Georgia, USA, from where the production of over a dozen brands, including Massey Ferguson tractors, GLEANER combines and WHITE planters is managed. There are regional offices in Coventry, England; Marktoberdorf, Germany; Sunshine, Australia; and Canoas, Brazil. Manufacturing takes place at five locations in the USA, six in Europe and two in South America. There are 20 warehouses spread across four continents that supply products and parts to farmers via a network of over 7,350 full-service dealers.

Source: www.agcocorp.com

which they are traded are imprecise and blurred. One firm's investments are another firm's borrowings, and capital appears in a number of different forms, including cash, equities, securities and bonds, as it circulates around the system. An important recent development is the proliferation of secondary financial products, such as futures, swaps and promissory notes, which have a derived monetary value. Speculative dealing in derivatives is a high-risk activity that generates large profits for those who are successful, but massive losses for those who are not.

The way in which financial products are traded is also complex. Some markets have a physical identity in the form of a stock or exchange building to which trading is restricted. Others exist as a set of individuals, working for institutions, who communicate with each other so as to buy and sell financial commodities or products. The level of regulation varies widely. Stock exchanges and bullion markets are tightly controlled. Other markets are informal or 'grey'.

Banking, in which arrangements are made to attract money from investors and to make loans to individuals, corporations, institutions and governments, lies at the traditional heart of the global financial system. Merchant banks and exchange houses play a central role in the circulation of money-making profits for their shareholders by charging fees for their services. Both the size and the complexity of this financial operation have increased in recent years, as the overall demand for funds for industry and economic development has risen, and as the sources of borrowing and destinations of investments have proliferated and diversified across the globe. Traditional forms of lending and borrowing through banks, however, are declining in relative importance, as alternative financial products, markets and institutions are being introduced.

The buying and selling of the currencies that are required for transactions in international goods and services take place in foreign exchange markets. Most foreign exchange transactions involve the buying or selling of US dollars, which function as an unofficial world currency. The fact that the value of currencies is liable to fluctuate, bringing gains or losses to those who hold them at the time, means that foreign exchange has become an important commodity in which institutions speculate. Substantial trade takes place in currencies, as dealers and brokers seek to make profits for their clients, and fees for themselves, through judicious buying and selling. Such currency trading has been helped by the recent internationalisation of domestic money, so that most of the principal currencies are traded across the world and around the clock. For example, the domestic currencies of the USA and Japan are freely traded in Europe as Eurodollars and Euroyens respectively in the rapidly growing Eurocurrency market.

Currency trading has increased in size and complexity in recent years through the development and exchange of swaps via the so-called swap market. Swaps are mutually beneficial arrangements between two parties to alter interest rates or currency exposure so as to avoid the need for unnecessary and perhaps unfavourable foreign exchange dealing. A further refinement is the growth of trading in agreements to buy currencies at specified prices at specified dates, again as a means of minimising risks and ensuring supplies. Exchanges such as the London International Financial Futures Exchange (LIFFE) facilitate this activity.

A separate and expanding set of activities is concerned with the issuing, buying and selling of financial securities. These are redeemable at a specified time, thus introducing an element of stability into what could otherwise be a dangerously volatile finance market. Financial securities typically take the form of bonds or promissory notes that are issued by borrowers and give the lender a specified income at an agreed maturity date and interest payments at regular intervals before that date. There are also numerous variants on fixed-rate bonds, together with other methods of securing debt obligations such as junk bonds. Banks manage the issue of new bonds and also trade in the so-called secondary market in which bonds that have already been issued are bought and sold.

A fourth set of global financial activities is concerned with stocks and shares. These securities are issued by brokers on behalf of governments and companies and are sold on the primary market to raise capital. They are then traded in a speculative fashion on the secondary market. Governments across the world apply strict regulations to the buying and selling of these items and dealing is restricted to stock and security exchanges (Plate 7.2). Although their traditional role was domestic, the leading exchanges increasingly issue and deal in stocks and shares for global companies and corporations.

A principal role of finance is to expedite the buying and selling of industrial raw materials. A long-established set of markets and institutions is concerned with mediating the exchange of commodities, including agricultural produce (e.g. cocoa, coffee, sugar, rubber, wool, soya meal), metals (copper, lead, zinc, nickel, aluminium), bullion (gold, silver) and oil. Some are sold for immediate delivery in the so-called spot market of which that in Rotterdam for oil is probably the best known. Most, however, are marketed via fixed-time, fixed-priced bills of exchange known as 'futures', which are used by corporations to ensure their supplies, to protect against commodity price rise and falls,

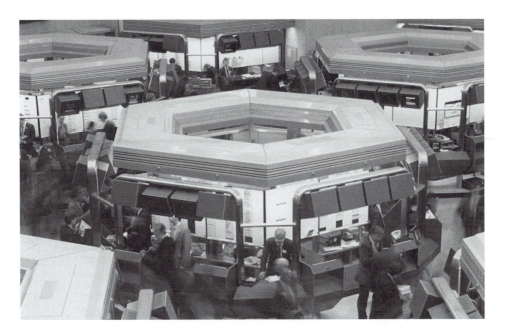

Plate 7.2 *The trading floor of the New York Stock Exchange*

or for speculation. They are traded in futures markets, examples of which include the London Metals Exchange, the London Gold Futures Market and the Chicago Mercantile Exchange.

The growth of the international service economy is associated with the rise of an advanced producer services sector. It includes insurance, accountancy, real estate, law, advertising, information technology support, research and development, public relations and management consultancy services (Table 7.2). A significant switch in the size and geographical scale of operation has occurred in recent years, such that the provision of key corporate services is now dominated by a small number of global companies (Beaverstock *et al.*, 1999). KPMG, PricewaterhouseCoopers, Ernst and Young and Arthur Anderson compete for a share of the global corporate accountancy market (Box 7.2). There are 11 major advertising companies (Box 7.3) and 16 that provide legal services on a global basis (Box 7.4). Estate agencies such as Goddard and Smith, Gooch and Wagstaff, Healey and Baker, Hillier Parker, May and Roden, Jones Lang Wootton, Knight Frank, Pepper Angliss, and Richard Ellis increasingly operate on a worldwide

Box 7.2 KPMG

KPMG is one of the leading global accountancy companies. It provides financial services to clients across industry sectors via 100,000 employees in over 150 countries. The company is increasingly multi-disciplinary, offering risk management, legal, taxation and corporate recovery services alongside traditional accountancy.

Source: www.kpmg.com

Box 7.3 Dentsu

Dentsu is a global advertising company with its head office in Tokyo. It employs 12,000 people and provides marketing, corporate communications, event promotion, strategy and branding services to corporate clients from 39 locations around the world. Regional head offices are maintained in New York, London, Paris and Düsseldorf and there are branch offices throughout Japan and in Ho Chi Minh City, Hanoi, Bangalore and Delhi.

Source: www.dentsu.com

basis. Global business is further facilitated by means of the organisation of employee services on an international basis. Examples include hotel accommodation (Hilton, Best Western, Holiday Inn), car hire (Hertz, Avis) and personal spending through credit and charge cards (MasterCard, Visa, American Express).

The concentration of command and control

As manufacturing, financial and service organisations have increased in size and have extended their spheres of operation across the globe, they have concentrated their headquarters functions in a small number of cities that have grown in world stature as a result. Twelve cities host the headquarters of 67 of the world's 100 largest corporations (Table 7.3). The top four cities alone account for the headquarters of 46. Tokyo is the principal choice, with 17 headquarters (including four of the top 20). New York and Paris are of roughly the same importance with 12 and 11 corporations respectively. The central role played by the core countries in the world economy is underlined by the fact that 11 of the top 12 cities are in Japan, the USA, the UK, France, Germany, the Netherlands or Switzerland. Seoul, with two headquarters, is the only developing world city in the top 12.

It is important to emphasise that it is the concentration of corporate power and not population size that contributes to world city status. Tokyo, New York, Osaka and Paris happen to have large populations and be centres of global corporate power, but there are no headquarters in any other of the world's largest 24 metropolitan areas as identified in Table 3.2. Munich, Amsterdam and Zurich are significant centres of corporate power, but are

Table 7.3 Headquarters of the world's 100 largest corporations, 2002

City	Metropolitan area population (000s)	Number of firms Top 100	Top 20
1 Tokyo	24,000	17	4
2 New York	16,640	12	1
3 Paris	9,624	11	1
4 Osaka	11,013	6	3
5 Detroit	3,788	4	2
6 London	7,640	3	1
7 Chicago	6,951	3	0
8 Munich	2,294	3	0
9 Amsterdam	1,144	2	0
10 Seoul	9,888	2	1
11 Frankfurt	3,687	2	0
12 Zurich	983	2	0

Sources: www.fortune.com; *Who Owns Whom?*; United Nations (2001).

only of medium size. Political status is similarly unimportant. Of the top 12 centres of corporate power, only Tokyo, London, Paris and Seoul are national capital cities (Table 7.3). New York hosts many corporate headquarters, but it is neither a national nor a state capital.

Financial institutions are similarly concentrated in a small number of world cities. They give rise to an elite group of centres that transact a disproportionate share of the world's financial business. The highest tier consists of supranational financial centres that are headquarters for a large number of internationally active banks that are well connected to other centres throughout the world. They are places where large amounts of foreign financial assets and liabilities are managed, where foreign direct investment capital is supplied to the rest of the world, and from where the organisational and operating rules and regulations that govern internationally active financial institutions are established. New York, London, Tokyo and Paris are the leading global financial centres on account of the size and diversified nature of their financial dealings (Beaverstock *et al.*, 1999). They have a large number and variety of financial institutions, including the head offices of many of the largest banks (Table 4.3). Frankfurt, Hong Kong, Milan, Paris, San Francisco, Singapore and Zurich also perform a supranational role, but have more limited global importance. A second level comprises those international financial centres which host large numbers of foreign financial institutions and the headquarters of a small number of internationally active banks. They have only a limited capacity to influence events that affect global asset and liability management. These places play an important financial role within their regions, but have only small stock markets and engage in relatively little truly international financial business.

More detailed analysis shows both the extent to which global finance is managed and controlled from a small number of world cities and the subtle differences of specialisation that exist among them. The degree to which the world's stock market activities are concentrated in New York, Tokyo and London is remarkable and throws into sharp relief the pretensions of aspiring or 'wannabe' centres, such as Miami and Sydney, to be world cities (Short, 1996). The global capital market is orchestrated by at most ten leading exchanges, with New York being by far the biggest in terms of market value (Table 7.4). With a market capitalisation of US$11 billion it is significantly larger than Tokyo and London (US$2.2 billion), although it lists a similar number of companies. Its size reflects the strength of the US economy and in particular the value of the many large American transnational corporations whose stocks and shares are traded. The status of London as a global, as opposed to merely a domestic, financial centre is underlined by large number of overseas companies quoted on its stock exchange. This is ten times that of Tokyo, which remains mainly a domestic exchange. A high level of international orientation is similarly a feature of the smaller exchanges in Frankfurt (Deutsche Börse) and Switzerland. The extent to which the major cities in the core economies command and control the global capital market is emphasised by the lower rankings. Of the 20 leading stock exchanges, only Hong Kong (ranked 10), Johannesburg (17) and Seoul (19) are in the developing world.

The concentration of corporate and financial activities gives rise to, and is supported by, a related geography of producer service organisations. Firms that supply advanced professional services are concentrated in world cities, where they support, and in turn are supported by, the major transnational corporations and finance houses. New York and London are the leading producers and exporters of international accountancy, advertising, management consultancy, legal and business services (Sassen, 1994).

Table 7.4 *International stock market comparisons, 2001*

	Market capitalisation***	Companies with shares listed		
		Domestic	Foreign	Total
New York	11,026,586	1,939	481	2,420
NASDAQ*	2,739,674	3,818	445	4,263
Tokyo	2,264,528	2,055	41	2,096
London	2,164,716	1,923	409	2,332
Euronext**	1,843,528	1,132	n/a	n/a
Deutsche Börse	1,071,748	748	225	973
Toronto	611,492	1,201	38	1,239
Italy	527,467	288	8	296
Swiss Exchange	527,374	283	149	432
Hong Kong	506,072	857	10	957

Source: www.world-exchanges.org

Notes:

* National Association of Securities Dealers Automated Quotations (a USA based electronic market).
** the Paris, Amsterdam and Brussels exchanges.
*** US$ millions.

The distinctive role played by world cities in the global economy is apparent when comparisons are drawn with principal domestic business centres. Lower order places may provide a wide range of corporate and advanced services, but this is for local rather than world consumption. Such differences in orientation are especially marked in the USA, where all the major cities, with the exception of New York, function principally as regional service centres. There is a similar contrast in the UK between London, which is a world city, and the regional centres of Birmingham, Manchester, Leeds, Edinburgh and Cardiff, which have little or no global importance.

Business information and decision-making

The key individuals and institutions of international production, finance, services and government concentrate in world cities because these are the best places from which to direct global activities. A complex set of location factors is involved, including access to information, economies of scale, attractions of prestige locations and exceptional global accessibility. Individual activities respond to these attractions in different ways. The fact that they support each other means that the benefits of world city location are cumulative and self-reinforcing.

World cities are the favoured locations for the headquarters of transnational corporations because they offer unparalleled access to business information. The principal responsibility of global corporate executives is 'orientation', which involves determining the nature and course of survival and growth of their organisations (Thorngren, 1970). This strategic task necessitates long-term scanning of socio-economic environments so as to identify and evaluate the factors that will affect the future operation of the business. The problem is then to choose a course of action that will minimise the threats to the firm and maximise the prospects for continued profitable operation. The aim of orientation is to ensure that the firm evolves in such a way as to be optimally placed in the future to take account of advantageous trading conditions. A wide range of influences needs to be considered, including raw material supply, the behaviour of markets, the activities of affiliates and competitors, developments in science and technology, the availability and cost of finance and labour, and government policies. Each is surrounded with uncertainty since it is determined by external and contextual factors that lie outside and beyond the control of the firm.

Corporate orientation is a risky activity and the stakes are correspondingly high. It involves making informed choices on new, unpredictable and non-standard problems about which there are many unknowns. Effective decision-making requires good quality business intelligence about the future environment in which the firm will operate. Of particular value is information or informed opinion on events such as stock market trends, fluctuations in currency values, wars, political upheavals and changes of government that can influence and affect the level of global business. Intelligence of this type cannot be generated within an organisation since it is to do with the behaviour of competitors, partners, politicians and states. It is most easily and reliably gleaned through regular face-to-face interaction with presidents, chief executives, directors and board members of global business, finance and producer services organisations. Members and officials of national and international governments who are similarly engaged in strategic planning and orientation for their own organisations also possess information which is of value in corporate orientation, as do research scientists who are engaged in work on new industrial processes and products.

The suppliers of advanced producer services are similarly drawn to world cities because of the access they afford to clients, collaborators and competitors. The tasks of arranging loans, financing capital projects and underwriting risk in a global economy involve detailed and confidential negotiations with a wide range of service and government organisations and it is of benefit collectively if the key representatives and decision makers are concentrated in the same place (Plate 7.3). It is also important for purposes of prestige for banks, insurance brokers, commodity traders and shipping companies to have a presence in the cities that are at the centre of world business and to be close to key marketplaces. These factors explain the exceptional concentration of corporate services in the 'square mile' of the City of London (Figure 7.1). The overriding requirement is proximity to the key institutions of the Bank of England, the Stock Exchange, Lloyd's insurance market (Box 7.5) and the Baltic Exchange (Box 7.6).

Telecommunications could undermine the benefits of concentration, but the evidence is that the principal financial centres are gaining rather than losing in importance. For example, in December 1994 the Deutsche Bank announced that, because of the exceptional access to the market, it was to concentrate its investment banking business in London through

Box 7.5 Lloyd's of London

The distinctive hi-tech Lloyd's building in the City of London houses the world's leading insurance market. It provides specialist insurance services to over 120 countries and handles the insurance business of 91 per cent of FTSE 100 companies and 93 per cent of Dow Jones companies.

Lloyd's consists of members or 'names' who provide capital to back up insurance risks. They are organised into 139 syndicates, run by managing agents, that specialise in particular types of insurance. Accredited brokers from 150 firms place risk in the Lloyd's market on behalf of clients by direct negotiation with agents. Proximity to Lloyd's is therefore essential and is a major reason for the high concentration of insurance companies in the City.

Source: www.lloyds.com

Plate 7.3 *The marketplace for insurance services: inside One Lime Street, the hi-tech home designed by Richard Rogers for Lloyd's of London*

The Baltic Exchange is an international self-regulated maritime market that provides its members with premises, facilities and agreed rules of conduct to enable them to undertake shipping and commodity business. It handles around 50 per cent of the world's tanker chartering business and 30–40 per cent of dry bulk chartering. It is also the largest market for the sale and purchase of ships, dealing in about half the world's new and second-hand tonnage. Fewer than 10 per cent of the deals initiated on the Baltic involve a British shipowner, importer, exporter or crew.

The Baltic has over 700 corporate and 1,500 individual members of some 45 different nationalities. Some are chartering agents who represent the merchants who have cargoes to move around the world. Others are brokers who act on behalf of the shipowners. The concentration of these representatives in a single building enables complex deals to be negotiated on a face-to-face basis, thus ensuring minimum delay and the best possible mutually agreed terms. The Exchange publishes daily market indices based upon data from broking houses around the world, thus establishing starting points for individual contract negotiations.

Source: www.balticexchange.com

Morgan Grenville, the British merchant bank, rather than push Frankfurt as the hub of its global corporate finance, share trading and derivatives operations.

Decision makers benefit collectively from living and working in close geographical proximity and from the resulting opportunities for generating and accessing business information. Personal contact enables the characteristics of associates, partners and competitors to be scrutinised and assessed and contributes to the building of 'confidence', reflected in the motto of the Baltic Exchange that 'our word is our bond', which is an essential prerequisite for successful business. It is helped by the fact that top executives are invariably drawn from the same narrow social backgrounds and have similar values and business ethics. Financial institutions in the City of London are especially in-bred; the majority of the employees are white males who were educated at public school and Oxbridge (Bowen, 1986).

The advantage of a world city location is the exceptional opportunity it provides for face-to-face contact with collaborators and competitors. Networking and corporate diplomacy take place at pre-arranged business meetings and conferences and also when executives meet informally in

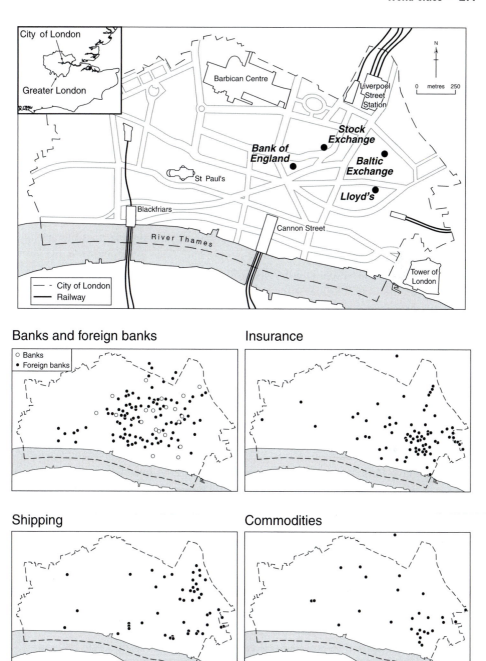

Figure 7.1 *Corporate services in the City of London*
Source: based upon maps in Clarke, 2001.

clubs, restaurants and in hospitality suites at major sporting and cultural events. Corporate entertaining is an important medium through which business relationships are established and from which information, intelligence and contacts may arise. The existence of a critical mass of senior executives and the scarce and high-quality business information that they command is the principal external economy that is provided by world cities. In place of the manufacture of goods, such places are now primary centres for the production of information.

Increasing global reach

The growth and concentration of global capital is made possible and is sustained by complex flows of people and ideas. Information is the key business resource and it is by controlling the flow of instructions, ideas and data to regional offices, branch plants, affiliates and subsidiaries that headquarters personnel are able to manage their global empires. Recent developments in telecommunications and transport facilitate and underpin this command and coordination function (Akwule, 1992). They permit what were once separate and dispersed economic activities to become integrated and concentrated functions. Together they have eradicated traditional barriers to interaction, such that time and space have collapsed to a point. Such points are world cities.

The rise of global financial centres has largely been made possible by telecommunications (Box 7.7). London, New York, Tokyo and Paris have long been of regional importance, but the introduction of international telecommunications and computer systems enabled them to develop global banking and financial trading functions. Of particular significance is the interlinkage of dealing rooms through dedicated and secure telecommunications networks, so that currencies, stocks, shares, commodity prices and volumes in every significant business centre are simultaneously displayed on computer screens around the world (Thrift, 2002). On this basis, 24-hour global trading can be conducted at the touch of a button, with financial settlements being made by electronic funds transfer (Leyshon and Thrift, 1997). Computers are commonly programmed to buy and sell automatically if prices fluctuate beyond specified limits. London is especially well placed to dominate global financial markets, as it lies between the Eastern Standard and Far Eastern time zones. Telecommunications enable a London-based dealer who starts work at 0600 hours to catch the end of trading on the Tokyo

Box 7.7 Global telecommunications

Telegraphy and telephony are nineteenth-century inventions, but it is only in the last 40 years that telecommunications have become media of instantaneous and low-cost worldwide communication. Global linkage became possible when the first intercontinental telephone cable was laid across the Atlantic in 1956. The potential of satellite communication became apparent in 1957 following the launch of Sputnik 1. It was facilitated in 1964 with the establishment of INTELSAT, an international consortium, led by the USA, which has created a network of communications satellites that gives worldwide coverage.

Global communication was enhanced in 1988 with the introduction of fibre-optic cable systems that have vastly increased transmission capacity. The most recent innovation is the development of an integrated services digital network that allows customers to send and receive high-speed, high-quality voice, data, image and text, or any combination, over public telephone lines. International ISDN links were first established in 1989 between the UK and France, the USA and Japan, while in 1990 service began to Australia, the Far East, Scandinavia and Germany.

The development and introduction of such technologies were helped by the recent deregulation of telecommunications industries across the world. The most important was the abolition in 1984 of the American Telephone and Telegraph (AT&T) monopoly which created new opportunities for competitors. In Great Britain, the privatisation of British Telecommunications and the entry into the market of the Mercury Corporation were similar processes. In 1987 the Japanese government licensed three new companies to compete with Nippon Telegraph and Telephone. The deregulation of the telecommunications industry was important because telephony is the leading form of telecommunications at the international scale. The major consequence was a reduction in long-distance call rates from which international voice and data transmission services benefited the most.

exchange, to trade all day in London, and then to conduct business for several hours in New York.

It is important to emphasise that telecommunications create and enhance rather than erode world city functions. Although telecommunications enable interaction to take place without participants travelling to central meeting places, they are not appropriate media through which to conduct the types of 'orientation' meeting that take place in world cities. The purpose of such meetings is to evaluate options, to negotiate deals and to take decisions. They typically involve top-level personnel and their advisors. It is essential for participants to be present in person since the aim is to float ideas, to gauge reactions, to cajole, to persuade and to decide. None of these activities can adequately be performed remotely. The face-to-face activities that take place in boardrooms and on the dealing floors in major financial centres have not been, and are unlikely to be, made obsolete by new technology; rather, technology has extended the global reach of those who transact such business and so has reinforced the status of world cities.

The pre-eminence of world cities is maintained and enhanced by the way in which new communications technologies are introduced. World cities benefit most from advances in telecommunications because, as established locations for global business, they are the places that first receive and derive the advantages of new services and applications. Telecommunications, in common with many innovations, diffuse hierarchically through urban systems. They are initially made available in major cities, where traffic levels, revenue and profits are greatest, and they subsequently spread down and outwards to successively lower-order centres. Early adopters gain the greatest locational advantages. They benefit in two main ways: first, by becoming more easily accessible to each other and, second, by experiencing a reduction in their communication costs as a result of the lower tariffs that are commonly associated with new technologies. Comparative advantage is compounded by the high level and frequency of innovation that has characterised the telecommunications sector in recent years. The latest transmission and terminal technologies are introduced before their predecessors have diffused completely through the urban system. Today, the complex world of global telecommunications includes systems that have been in use for many decades alongside other innovations that are barely ten years old. World cities are the principal beneficiaries since they are typically several technologies ahead of competing lower-order centres.

An integrated world economy dominated by world cities is largely the creation of global telecommunications, as there have been few comparable improvements in business travel. The network of air connections has expanded rapidly in recent years, but the last major technological advance in travel was the introduction, as long ago as 1976, of supersonic Concorde aircraft on the north Atlantic routes. This reduced the flying time between London Heathrow and New York's John F. Kennedy airports from eight hours by subsonic services, to four hours. Travel times on very long-haul routes, including those between London, New York and the principal Asian business centres, have recently been reduced because of the introduction of non-stop flights, but otherwise the airline industry has been for the last 30 years on a performance plateau. Travel times between city centres have probably risen because of road traffic congestion and the time now required for check-in, security clearance and baggage reclaim. Regular long-haul business travel is tolerated by most, but relished by few, to the relative advantage of telecommunications links. Further significant reductions in air travel times are not envisaged until hypersonic engine technology becomes a commercial reality.

The central importance of telecommunications in supporting world city functions is reflected in the changing design and construction of office buildings. Global financial companies need vast open spaces to accommodate their dealing rooms and trading floors, together with high ceilings to carry the cabling for telephones and computers and the ducting for air conditioning. Such requirements are met by the construction of dense deep-plan buildings that cover the whole of the site. They typically feature central courtyards and atria that bring daylight into their centres. Such 'groundscrapers' contrast markedly with the 1960s thin-clad high-rise office blocks with their inadequate servicing, forests of internal columns and low ceilings (Williams, 1992). Their proliferation in the urban landscape provides powerful visual evidence of the fundamental importance of telecommunications to the command and control functions of world cities.

Representing and promoting places as world cities

Many civic leaders around the globe aspire to world city status for their cities because of the power, wealth and prestige this position brings. World city functions, in comparison to manufacturing industry, employ large numbers of people at low environmental costs. They produce significant incomes to the city, and the salaries that they pay generate powerful multiplier effects within the national and urban economy. The City of London, for example, produces massive 'invisible' exports through the global selling of financial and insurance products and contributes about 5 per cent of the GDP of the UK (Clarke, 2001). World city functions are increasing in importance worldwide as the global economy develops. They are footloose, independent of natural resources and can in theory locate anywhere that has appropriate office space and global communications connections. There is keen competition among wannabe world cities, amounting at times to 'place wars' (Haider, 1992), to grab the largest possible share of corporate, financial and advanced producer service activities.

The promotion of places as world cities is a major industry that involves image creation and manipulation, marketing and merchandising. It extends across a wide range of areas, including the media, heritage, sport, recreation, tourism and aspects of the urban economy. The aim is to give the city a clear positive image or identity so that it will stand out from other places. It is to present the city in the most flattering light and to boost its appeal so that it will attract global attention and investment. Images are constructions that shape perceptions and define the ways in which cities are viewed and responded to. They create a landscape of signs and symbols that has replaced use as the basis for consumption (O'Connor, 1998).

The clarity of urban images depends upon the ease with which places can be 'read'. Legibility is affected by the character of the city's physical environment, so places with exceptional natural attributes, such as Hong Kong, Rio de Janeiro and Sydney, are easily represented. It is enhanced by the presence of architectural qualities and heritage features that generate identity and appeal. Images may be constructed collectively through representations in literature, art, music and the media, as well as by conscious creation by image consultants working in video, film and the Internet. They may have a basis in reality, but in many cases they are little more than a marketing fiction.

Strong positive identities exist for some cities, without the need for deliberate manipulation, through popular rhetoric. For example, lines in songs such as 'Wonderful, wonderful Copenhagen', 'Vienna, my city of dreams', 'I love Paris in the springtime' and 'I left my heart in San Francisco' give these cities identity and appeal in the same way as do Canaletto's paintings of Venice and the Impressionists' depictions of Paris. It is difficult to find similar examples for Almaty, Ho Chi Minh City and Minsk. Representing and promoting former centres of manufacturing as world cities is difficult because of negative imagery associated with the industrial past. Identity is facilitated by the presence of iconic structures and artefacts, which commonly have a disproportionate importance in shaping the urban image. London, Paris, Athens, Sydney, New York and Washington are instantly recognisable from pictures of Tower Bridge, the Eiffel Tower, the Parthenon, the Sydney Opera House, the Statue of Liberty and the White House respectively. The converse is also true. Zurich,

São Paulo and Melbourne have aspirations to be world cities, but, lacking icons of equivalent importance, suffer from low visibility and recognition.

The attraction of footloose corporate and global service functions is influenced by the quality of office space, airports and cultural outlets, as well as by perceptions of what a city is like and has to offer. Global organisations require and expect the best and it is essential that civic authorities create a very high quality physical environment. One consequence is the competition to have the world's tallest building, an accolade that is currently held by the Petronas Towers in Kuala Lumpur. Another is the keenness to add prestige by commissioning signature buildings from big-name international architects, including Richard Rogers, Norman Foster, James Stirling, Michael Graves, Arata Isozaki, I. M. Pei, Aldo Rossi and Philip Johnson (Plate 7.4). Such buildings are a powerful growth rhetoric and make clear statements about the dynamism of the city and its determination to attract footloose global capital.

Cities promote themselves and advance their claims for world recognition more generally by hosting global spectaculars, such as major conferences and sports events. The Olympic Games are the big prize and the competition to hold this four-yearly event provides unparalleled opportunities for cities to monopolise world attention (Table 7.5). The host and candidate cities between 1992 and 2004 are all to an extent wannabe world cities (Short and Kim, 1999: 100). Kudos is also associated with the holding of major United Nations conferences and lesser sporting events, such as the football World Cup, the World Student Games and the Commonwealth Games.

Table 7.5 *Host and bid cities for the Olympic Games, 1988–2008*

Year	Host city	Bid cities
1988	Seoul	Nagoya
1992	Barcelona	Amsterdam, Belgrade, Birmingham, Brisbane, Paris
1996	Atlanta	Athens, Belgrade, Manchester, Melbourne, Toronto
2000	Sydney	Beijing, Berlin, Istanbul, Manchester
2004	Athens	Buenos Aires, Cape Town, Rome, Stockholm
2008	Beijing	Istanbul, Osaka, Paris, Toronto

Source: www.olympic.org

Plate 7.4 *Signature buildings: the head office of the Hong Kong and Shanghai Bank Corporation, in Hong Kong, designed by Norman Foster*

Important questions, however, surround the extent to which wannabes can achieve world city status through image creation and promotion and whether the prize is worth the effort involved. The world city attributes of Tokyo, London, New York and Paris have accumulated over many centuries and form a critical mass in each place that is unlikely to break up or be replicated elsewhere. Their roles are built around key exchanges and marketplaces that are deeply embedded within the business cultures of these cities. The biggest threat to existing world cities is more likely to be erosion of functions from within rather than competition from wannabes, as electronic exchanges, such as the NASDAQ, replace place-based markets (Plate 7.5).

Plate 7.5 *Electronic trading is likely to replace face-to-face exchanges, thus undermining the viability of world cities*

Promoting places as world cities is an expensive zero-sum game in which the only clear winners are marketing analysts and image consultants. The benefits of hosting world spectacles are uncertain. The Olympic Games in Sydney in 2000 were a major success and significantly enhanced the city's image and appeal. The spin-offs from Atlanta in 1996 were disappointing. Montreal is still paying off debts incurred by hosting the Games in 1976.

Tokyo as an example of a world city

The roles played by the institutions of global capitalism, their geographical concentration, and transport and telecommunications are well illustrated by the rise of Tokyo as a world city. In contrast to London and New York, Tokyo was a late entrant into the world city league. Up to the 1930s Japan was a bipolar country, with Tokyo and Osaka competing for national pre-eminence. Significant investments were made at the time in colonies and other territories in eastern Asia and it was imperialist aspirations that were responsible for Japan entering the Second World War. Defeat and subsequent occupation gave rise to a geographical restructuring in the form of a concentration of economic activities around Tokyo, which laid the foundations for national and subsequently world city status. During and immediately after the war many Osaka-based industrial corporations relocated their headquarters to Tokyo as the centre of political power, as did many non-Tokyo banks. A similar trend occurred in higher education and research, such that the University of Tokyo emerged as the dominant educational institution (Markusen and Gwiasda, 1994). The fact that research-led microelectronics is the leading sector in the national economy meant that Tokyo became the principal manufacturing centre in Japan.

Its continued success as an industrial centre owes much to the development and exploitation of systems of flexible production, which enable manufacturers to switch rapidly from one product to another in response to changes in domestic or world markets. Such systems extend beyond the availability of flexible machinery and skilled labour to include networks that bind research and development specialists, manufacturers, distributors, government agencies and service suppliers in a highly responsive production arrangement (Fujita, 1991). Flexible production is facilitated by, and in turn supports, a distinctive industrial structure in which most manufacturing is undertaken in small factories. Most large plants with over 300 workers have moved out of Tokyo in recent years and the number of small plants has increased proportionately. 'If the total number of plants in 1960 is indexed at 100, the number of large plants fell to 53.6 by 1983, but the number of small plants with less than 299 workers rose sharply to 185.8' (ibid.: 275). Presently, some 84 per cent of plants employ between four and 19 workers. The concentration of small flexible producers gives Tokyo a special character as a manufacturing city. It is a centre of innovation – a place where new industries and hybrid industries originate.

Tokyo became a world city as a consequence of the transnationalisation of Japanese commercial capital that was generated through manufacturing. In the early post-war period Japan's foreign investments were aimed primarily at securing raw materials, such as oil, wood and pulp, for domestic industrial production. After about 1970 capital exports were liberalised and gave rise to two distinct waves of capital investment. The first was in factories in nearby Asian and Pacific countries with low labour costs, in order to manufacture electronic components and semiconductors. The second was in advanced western economies, including the USA, in the form of plants through which Japanese electrical goods and cars could be imported and assembled.

As globalisation proceeded, Tokyo became the principal centre in Japan for international trade and finance. A rapid increase in the number of corporate headquarters in the city occurred in the 1980s, as corporations sought to locate in the central business district for symbolic reasons, to gain international competitiveness, to have access to information sources and to deal with international trading and transactions. The wealth that they generated was recycled overseas through foreign securities and currency markets by Japanese banks, by foreign financial institutions that were attracted to Tokyo and by branches of Japanese banks in other world cities, principally London and New York. The presence of Japanese corporations in the global economy attracted foreign direct investment and employment by American and European companies, most of which was concentrated in Tokyo. Some 66 per cent of the 70,000 foreigners in Japan in 1988 lived in the city. Tokyo's status as a world city derives from its concentration of corporate head offices, financial institutions and supporting producer service organisations that exceeds that of any centre in Asia.

Although the translation of domestic economic power into global influence is a process that occurred in most world cities, Tokyo is different to its principal rivals in several important respects

(Machimura, 1992). First, it is mainly a world economic centre that is unsupported by a political or military hegenomy of the state. Second, it is heavily dependent upon global communications systems because it is physically far away from the western countries that are the traditional centres of the world economy. Advances in telecommunications and transport, assisted by the success of the Japanese economy in electronics, were indispensable for it to overcome its geographical disadvantage. Third, Tokyo was for many years closed to the beneficial influx of foreign immigrants. For Machimura (1992), New York and London are world cities because of the political, military and cultural hegemony of the state. Tokyo is a world city because of its global network of economic activities.

Conclusion

The cities that are identified and analysed in this chapter occupy the top tier in the global urban hierarchy. They are distinguished not by their size or status as national capitals, but by their specialisms and the roles that they perform. Such is the degree of concentration of the headquarters of transnational corporations, the head offices of banks, finance dealers and the supporting producer service organisations that they function as command and control points for global capitalism. Their true status and the way in which this is conveyed in the form of a descriptive label are matters of legitimate and unresolved debate. What is clear, however, is that the individuals and institutions located within them exert a disproportionate and, according to some analysts, decisive influence on the shape and structure of the contemporary urban world.

Many places lay claim to world city status, but only New York, Tokyo, London and Paris have the critical mass of corporate, financial and service functions to satisfy the criteria for inclusion in this elite group. Other, lesser, centres have some global functions. Still others are wannabes that represent and promote themselves as world cities through image creation, marketing and the hosting of spectacles. Heads of corporations and governments, and their key staff, cluster in world cities so as to generate and to benefit collectively from information and to further the interests and activities of their businesses by being at the centre of world markets. They manipulate and play these markets via telecommunications and computer services that give low cost and instantaneous global reach. Dominant cities are associated with the world systems of the past, as Chase-Dunn (1985) has emphasised, but the world cities today are the first to perform a global role. Their emergence reflects and underlines the extent to which capitalism, and hence global urban development, is shaped by a minority of decision makers working out of a small number of key locations.

World cities dominate the settlement hierarchy in a highly developed urban world. They are the control points for a capitalist system that has helped to concentrate large and growing numbers of people in urban places. Many live in rapidly expanding mega-cities, especially in the developing world. Although the present pattern appears to be viable, although unjustly structured, there are many concerns as to its future. How it is likely to evolve and whether it is sustainable are topics that are addressed in the final chapter.

Recommended reading

Abu-Lughod, J. L. (1999) *New York, Los Angeles, Chicago: America's Global Cities*, Minnesota: University of Minnesota Press. An analysis of the development and role of three of the places in the USA which have claims to world city status.

Clarke, W. M. (2001) *How the City of London Works*, London: Sweet and Maxwell. This is a simple, straightforward and non-technical introduction to the financial markets of the City of London. It describes the key institutions and exchanges and explains how they work and the ways in which they are regulated.

King, A. D. (1990) *Global Cities: Post-imperialism and the Internationalisation of London*, London: Routledge. This book combines a scholarly analysis of the development of the world economy and the role of London as a colonial, imperial and world city. The treatment of the social and physical consequences of London's world city status is comprehensive and highly detailed.

Knox, P. L. and Taylor, P. J. (eds) (1994) *World Cities in a World System*, Cambridge: Cambridge University Press. This book comprises a set of research papers, originally presented at an international conference, on the world city hypothesis and the world system. All aspects of the current debate on world cities and their role in the global economy are discussed.

Sassen, S. (1994) *Cities in a World Economy*, London: Pine Forge. A sociological examination of the urban impact of economic globalisation and the rise and role of world cities.

Sassen, S. (2000) *The Global City: New York, London, Tokyo*, Princeton: Princeton University Press. A detailed and incisive analysis of the development and role of the three major world cities.

Short, J. R. and Kim, Y. (1999) *Globalisation and the City*, Harlow: Pearson. An introductory text that examines the urban impact of globalisation and the role of cities in globalisation. There are relevant chapters in part 2 on economic globalisation, the global urban system and world cities.

Key web sites

www.lboro.ac.uk/gawc/   The home site for the University of Loughborough's *Globalisation and World Cities* research and study project. This superb resource provides access to several data sets and numerous research papers on world cities, many of which have been published in refereed journals. It further provides links to web sites which give information on all the major world cities.

www.world-exchanges.org   A site that provides data on the world's stock markets.

Topics for discussion

1 Critically evaluate the concept of the world city.

2 With reference to specific examples, outline and account for the characteristics of world cities.

3 To what extent are world cities 'dual cities'?

4 'Big cities but not world cities'. Discuss this assessment of the contemporary status of New York, London, Tokyo and Paris.

5 What factors explain the growth of 'world cities'?

8 The future urban world

By the end of this chapter you should:

- be aware of likely future changes in the growth and distribution of urban populations;
- be familiar with the concept of sustainability and the debates which surround it;
- understand the issues of sustainability which surround the viability of cities;
- be aware of, and be able to evaluate, initiatives to enhance sustainability.

Introduction

It is appropriate in a book that adopts a historical approach and a global perspective to conclude with a brief consideration of how the urban world may evolve in the near future. Forecasting population levels, distributions and socio-economic conditions in a single country is a difficult enough task and attempts to undertake this sort of exercise at the global scale can only yield predictions that are little more than guesstimates. This chapter is therefore grounded in speculation rather than in detailed analysis. What is clear, however, because they are products of long-term and deep-seated processes that have yet to run their course, is that urban growth and urbanisation will lead to further significant urban development. Most will be in those parts of the world that are presently classified as developing. Even over the next quarter of a century the changes that are expected are staggering. Extrapolation of current trends suggests that the number of people who presently (2003) live in urban places is likely to rise by 50 per cent by the year 2025.

The scale of urban development that these figures imply raises important questions as to whether such a geographical pattern can be supported. It is difficult to imagine a world with significantly more urban residents than today and it is important to focus on issues of maintenance and sustainability. Cities are elements in global social, economic and environmental systems that are both vulnerable and fragile. Although they represent a highly efficient use of space and provide unrivalled opportunities for production and social interaction, they consume prodigious amounts of finite resources, far more than a rural population of equivalent size. There are grave doubts as to whether future cities can be sustained in economic terms and how they can generate and distribute sufficient wealth to support their residents at acceptable standards of living. A parallel concern is with the ecological implications of further urban development. Cities are widely seen as being parasitic, in that they draw air and water from the natural environment, but generate large quantities of pollution and waste. Little is reused and recycled. Many argue that their emissions are progressively destroying the global environmental systems upon which life on the planet, and hence their own viability, depends. The prospect for further massive urban development necessarily focuses attention upon the implications for the environment and whether the urban future is sustainable.

Such questions raise a further set of issues concerning the need for regulation today so as to ensure the continuation of cities into the future. If urban life is to be sustained, then steps must be taken now so as to prevent further damage to the environment and bequeath adequate resources to succeeding generations. Action is required at the global scale to cut harmful emissions and prevent indiscriminate and unnecessary exploitation of scarce resources. Within national boundaries, there is a need to deal with local sustainability issues, including urban servicing and waste disposal. Agendas for such intervention are presently emerging, although there are more high-level statements of intent than examples of concerted and effective action. The urban future is likely to depend as much upon the success of international agencies and governments in shaping urban development as it is upon the unregulated growth and redistribution of the population.

The urban future

The direction and scale of contemporary urban growth and urbanisation point to the emergence by 2025 of an urban world that will bear little resemblance to the urban present. This much is certain, but filling in the detail by country and by region is problematical because of the deficiencies of the data, which have been emphasised throughout this book and which are addressed in the Appendix. A related difficulty is methodological and concerns the interpretation of past trends and the ways in which they are used in forecasting. Many past predictions of the urban present have proved to be wide of the mark. For example, continuing urbanisation in the USA, resulting in a proliferation of megalopolises by the end of last century, was envisaged in 1967 by Kahn and Weiner in their book *The Year 2000*. This pattern, however, failed to materialise and the reality was massive deconcentration of population as a result of powerful and cumulative processes of counterurbanisation. The course of future urban development is uncertain. It is too sensitive to economic, social and environmental change to predict more than one or two decades ahead.

An example of the very different pictures that can emerge at the world scale is provided by the projections for city populations in the year 2000 that were published by the United Nations between 1973 and 1984 (Table 8.1). The projected population of Mexico City in the year 2000 was 31.6 million in the Population Division's 1973–5 assessment, but was much lower at 25.8 million in the 1984–5 forecast. Similarly, the 1973–5 projection for the population of Seoul in 2000 was 18.7 million, but this had been revised downwards to 13.8 million in 1984–5. The population of Mexico City and Seoul in 2000 was 18.1 million and 10.0 million respectively. The scale of revision by the leading statistical agency underlines the innate difficulties that are involved in long-term urban forecasting.

The most recent estimate by the United Nations (2001a) is that, by the year 2025, there will be some 4.6 billion people out of a world

Table 8.1 *Examples of changing projections for city populations by the year 2000*

City	UN projection for population				
	1973–5	1978	1982	1984–5	2000
Mexico City	31.6	31.0	27.6	25.8	18.1
São Paulo	26.0	25.8	21.5	24.0	17.8
Calcutta	19.7	16.4	15.9	16.5	12.9
Rio de Janeiro	19.4	19.0	14.2	13.3	10.6
Shanghai	19.2	23.7	25.9	14.3	12.9
Mumbai	19.1	16.8	16.3	16.0	18.1
Seoul	18.7	13.7	13.5	13.8	10.0

Sources: United Nations Population Division, Department of International Economic and Social Affairs; United Nations (2001a: Table B1).

population of some 7.9 billion living in urban places. This future urban population is roughly the same as the total population of the world today. Some 4.4 billion will be living in towns and cities in what are presently classified as developing countries. The population of urban places in China will be close to 800 million and in India is expected to be some 500 million (Figure 8.1).

Urban growth will be accompanied by increased urbanisation. Some 65 per cent of the world's population is expected to be urban by the year 2025. It follows from the analysis and discussion in Chapter 4 that this increase will occur principally because of the urbanisation of the population across large parts of Africa and Asia. These regions will be most radically affected by urban development, both urban growth and urbanisation, in the next quarter-century (Box 8.1).

The most striking feature of the predicted urban geography of the year 2025 is the uniformly high level of urban development in the Americas (Figure 8.2). The population of all of the principal countries of North, Central and South America is expected to be over 60 per cent urban and in most it will be in excess of 80 per cent. The Americas are presently highly urbanised, so this change represents a consolidation of existing patterns. Similarly high levels of urban development are anticipated in Australia, Japan, parts of the Middle East, northern Africa and most of Europe. Urbanisation levels in excess of 60 per cent are expected across the whole of Asia north of the Himalayas.

Box 8.1 Urban Africa in 2025

The distribution of population in Africa is not expected to become more urban than rural until 2025, by which time 50.4 per cent of the continent's 1.4 billion people will be living in urban places. Many countries, however, will remain predominantly rural, with the population of Burundi, Burkina Faso, Ethiopia, Malawi, Rwanda and Uganda being less than 30 per cent urban. There will, like today, be few major urban centres. Lagos, with a population of 23 million, and Cairo, with 14 million, will be the only places with more than ten million people. Some 43 other cities will have populations in excess of one million, eight of these being in South Africa and ten in countries along the Mediterranean coast. Primacy will characterise many domestic urban hierarchies. The distribution of urban centres will be grossly uneven, with large parts of the continental interior being remote from urban influences.

Source: United Nations (2002).

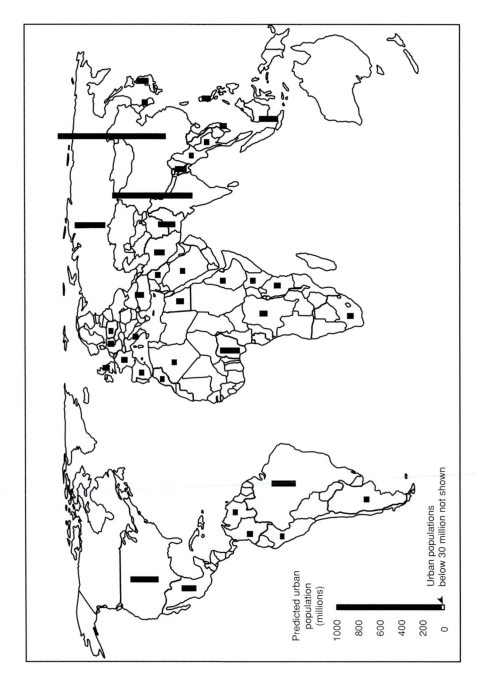

Figure 8.1 *Predicted urban populations, 2025*

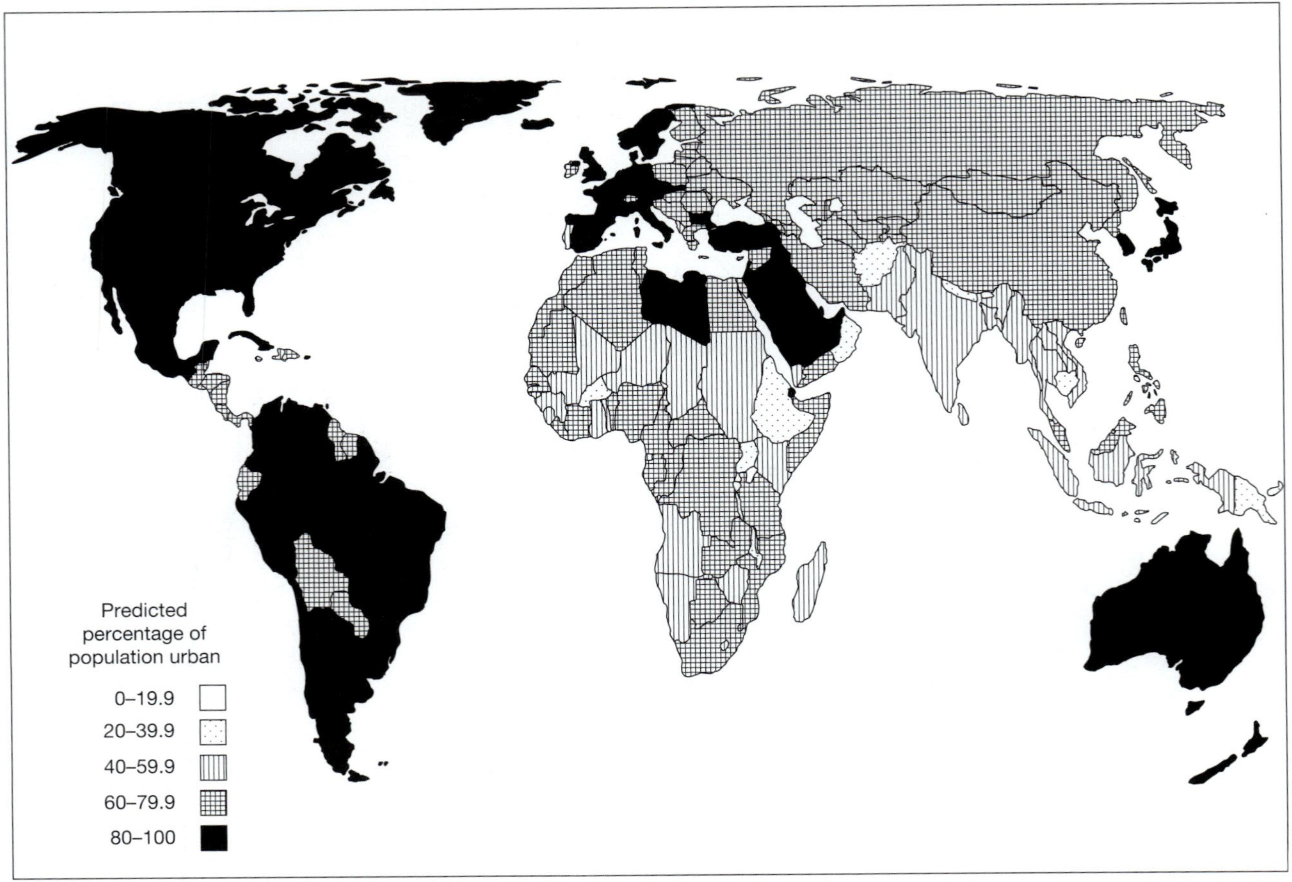

Figure 8.2 *Predicted percentage of population urban, 2025*

The United Nations forecast suggests that levels of urban development in Africa and southern Asia will be very much higher than today, but will vary considerably from country to country. Although the population in most countries will be more urban than rural, the proportion living in towns and cities in Burundi, Malawi, Rwanda, Ethiopia, Uganda, Burkina Faso, Afghanistan, Nepal, Bhutan, Cambodia, East Timor and Papua New Guinea is expected to be less than 40 per cent. Such countries have yet to go through the phase of rapid urbanisation that is a characteristic feature of the cycle of urban development (Figure 3.3). They will be the world's last remaining rural territories. At 51 per cent, India is expected to be only marginally more urban than rural in 2025.

If migration continues as it has in the past, and there is presently no suggestion that it will not, many who move from rural areas to urban places will go to the largest cities. The number of cities with populations of over 10 million is expected to rise from 24 in 2000 to more than 40 in 2025 (United Nations, 2001a). Strong patterns of primacy already exist in many African countries and these are likely to be reinforced rather than reduced by urbanisation and urban growth. An important consequence of continued primacy is the further polarisation of the settlement hierarchy. The problem will not so much be the excessive size of one city, but the smallness of most of the others. Even by 2025, in the least urbanised countries in Africa, there is expected to be a deficiency of urban centres that are large enough to support the variety of services needed to stimulate the commercialisation of agriculture, to meet the basic needs of the rural poor and to increase the productivity and income of the rural population. A widening of the gap between town and country is expected to be one of the major consequences of the urban transition in Africa over the next 30 years.

The urban geography of the developed world is likely to be very different. Here, urban populations are presently high and the principal shifts will take place among cities. Rather than a concentration in a small number of large cities, which is the current pattern, the population is expected to be more evenly spread across many smaller centres. Cities of about 200,000 are likely to be the most attractive, as they are large enough to sustain an acceptable range of services without the congestion and pollution that are associated with life in the mega-city. As the benefits of centrality and agglomeration lessen further, many people will be drawn to places which may be too small to merit designation as urban. The spatial structure of large cities is likely to be transformed by shifts in population and economic activity. Central area densities are likely to fall significantly as people and businesses move to the suburbs and beyond. Many of the areas which are vacated will become parkland and open space, so that cities will take on a 'doughnut' form. In the longer term, however, it is possible, as Table 3.3 suggests, that central area populations will rise as cities begin to reurbanise. Decentralisation of population at the local scale, and deconcentration at the national level, will significantly reduce urban/rural differences, thus producing a 'rurban' arrangement. The expected trends in different parts of the world will lead to a progressive inversion of the contemporary urban pattern at the global scale. A small city and rural orientation will increasingly characterise the landscape in developed economies, but strongly urban-centred mega-city societies will emerge and predominate in the developing world.

Issues of sustainability

Important questions concerning the viability of future urban developments are raised by these forecasts. It is difficult, given the current pressures on resources and the environment, to see how urban populations can increase further without some form of social, economic or ecological breakdown. How will the urban population be fed and what effects will mass concentration of population have upon the global environment and upon local ecosystems? The concept of sustainability is central to this debate. It is concerned with the support for urban concentrations and the need to take action today so as to ensure continuity and prosperity in the future (Box 8.2).

Sustainable development can be interpreted in a number of different ways – indeed, Pearce *et al.* (1989), in an appendix to their book *Blueprint for a Green Economy*, quote 24 separate definitions. The most widely accepted is that of the report of the World Commission on Environment and Development (WCED), also known as the Brundtland Commission, which suggested that sustainable development is that 'which meets the needs of the present without compromising the ability of future generations to meet their own needs' (1987: 43). This definition elevated the notion of sustainable development to the level of an operational concept, which embodied the principles, and values, which are desirable and necessary so as to deal effectively with the crisis of the environment and the development process. Emphasis was placed on the need for action today so as to provide for social, economic and ecological viability tomorrow.

Box 8.2 Sustainability

Haughton and Hunter (1994) argue that the concept of sustainability involves three major principles. The first is that of 'intergenerational equity' and concerns the legacy that is left to future generations. The success of cities in the future depends to a large extent upon the assets and resources that are available and it is therefore incumbent upon the current generation not to indulge in indiscriminate and wasteful consumption. A sustainable future requires that national capital assets of at least equal value to those of the present are passed on to succeeding generations.

A fair and equitable use of present resources is clearly necessary and this is enshrined in the second principle of 'social justice'. Some form of central control over access to, and use of, resources is implied. The fact that both resources and consumption are widely distributed and are interdependent means that such management must be at a broad scale. A third precondition for sustainable development is that of 'transfrontier responsibility', meaning that key issues such as pollution, waste disposal and climatic warming are not constrained by national or regional boundaries, but are essentially global in cause and consequence.

A general concern with sustainable development emerged as a key issue in the heightened environmental awareness of the late 1980s (Camagni et al., 2001). It gave rise to the emergence of 'green' parties in many developed countries in which the membership was committed to forcing environmental issues to the centre of the political debate. Evidence for global warming, the depletion of the ozone layer and the detrimental effect of acid rain highlighted the need for urgent action to prevent further environmental degradation. Debate on sustainability is, however, strongly infused with speculation and conjecture. Hard information is lacking and the significance of such scientific evidence that is available, as for example on the scale of ozone depletion and global warming, is much disputed. Particular difficulties surround the availability of data for developing countries, where environmental monitoring is in its infancy. Such problems do not of themselves negate discussion, but they mean that it must be conducted at a very general level. It follows that few clear conclusions can be reached that might assist urban planners and managers.

Against this background, Blowers (1993) identifies five fundamental goals that should guide all decisions concerning future development so as to ensure sustainability. The first concerns conservation and involves the need to ensure the supply of natural resources for present and future generations through the efficient use of land, less wasteful use of non-renewable resources, their replacement by renewable resources wherever possible, and the maintenance of biological diversity. The second concerns the use of physical resources and their impact on the land. It seeks to ensure that development and the use of the built environment are in harmony with the natural environment and that the relationship between the two is one of balance and mutual enhancement. A third goal is to prevent or reduce the processes that downgrade or pollute the environment and to promote the regenerative capacity of ecosystems. The final two goals are social and political in character. The aim of the fourth goal is to prevent any development that increases the gap between rich and poor and to encourage development that reduces social inequality. The final goal is to change attitudes, values and behaviour by encouraging increased participation in political decision-making and in initiating environmental improvements at all levels from the local community upwards.

Urban sustainability

A specific focus upon the sustainability of cities arises because it is at the urban level that many environmental problems originate and where they are experienced with greatest intensity. Cities exist at the expense of the environment, as Chapter 2 has shown, and have extensive supply and support footprints, which may be worldwide (Box 8.3). They extract far more from the environment than they return. Cities deplete energy, generate waste and create noise and congestion (Capello et al., 1999). They have an especially detrimental impact on the atmosphere, as air pollution is more wide-ranging and difficult to treat than is water or ground pollution (Box 8.4).

The concern for sustainability arises out of a recognition that urban development is a linear process: 'Food, fuels, construction materials, forest products and processed goods are imported into the city from somewhere, never mind where, and when they are finished with they are discarded, never mind how' (Giradet, 1990: 7). For example, it is estimated that only 15 to 20 per cent of urban-sourced nitrous oxide pollution falls inside the city, the rest contaminating rural areas often many kilometres away (Alcamo and Lubkert, 1990). Giradet points out that this linear system is profoundly different to nature's own circular metabolism in which every output is also an input that renews and sustains life. As the urbanisation level increases, this imbalance has

Box 8.3 London's ecological footprint

An extensive area of land is required to supply cities with food and timber products and to absorb the carbon dioxide that they generate. London's total footprint is estimated to be around 125 times its surface area of 158,000 hectares, or nearly 20 million hectares. With 12 per cent of the country's population, London requires the equivalent of Britain's entire productive land. In reality this land includes the wheat prairies of North America, the tea plantations of Assam, the forests of Scandinavia and the copper mines of Zambia.

London: 7,000,000 people
Surface area: 158,000 hectares
Area required for food production: 1.2 hectares per person: 8,400,000 hectares
Area required for wood products: 768,000 hectares
Area required for carbon dioxide absorption: 1.5 hectares per person: 10,500,000 hectares
Total footprint: 19,700,000 hectares
Britain's surface area: 24,400,000 hectares

Source: Gairdet (1998)

Box 8.4 Atmospheric pollution

Beyond critical concentrations, all pollutants discharged to the atmosphere are harmful to plants, animals and humans. The major pollutants in the urban atmosphere stem mainly from the burning of fossil fuels. They include:

Carbon dioxide, which prevents heat escaping from the planet and is a major influence on global temperatures.

Carbon monoxide, which interacts with other pollutants to produce photochemical smog and surface ozone.

Nitrogen oxide, which combines with sulphur dioxide to form the acid rain that has such serious detrimental effects on many ecosystems.

Sulphur dioxide, which contributes to acid rain and can cause bronchitis and diseases of the respiratory tract.

Volatile organic compounds, comprising a variety of hydrocarbons and other substances. These can combine with other pollutants to cause low-level ozone that dims sunlight and causes discomfort to the eyes and nose.

Particulate matter, which reduces visibility and carries toxic substances that are injurious to health.

massive implications for the well-being of the world's forests, soils, watercourses and atmosphere. Such degradation in turn threatens the cities, which contribute to it.

Some indication of the highly detrimental effect of cities is provided by measures of the levels of suspended particulate matter, such as soil, soot, smoke, metals and acids, which are found over cities as opposed to in adjacent rural areas. Goudie (1990) reports the average concentration of suspended particulate matter found in the commercial areas of a number of cities as: 400 micrograms per cubic metre in Calcutta; 170 in Madrid and Prague; 147 in Zagreb; 43 in Tokyo; and 24 in London and Brussels. These values compare with rural concentrations of less than 10 micrograms per cubic metre in rural areas.

An estimated 1.4 billion urban residents worldwide are exposed to averages for suspended particulate matter or sulphur dioxide (or both) that are higher than the levels recommended by the World Health Organization. Research reported by Hardoy et al. (1992: 79) underlines the severity of the pollution that hangs over many developing world cities. For example, in Shanghai there are seven power stations, eight steel works, 8,000 industrial boilers, 1,000 kilns, 15,000 restaurant stoves and one million cooking stoves, most using coal with a high sulphur content. In 1991 the annual average concentration of sulphur dioxide in the urban core was more than twice the recommended level. The annual average for total suspended solids was more than four times that recommended. A similar situation exists in São Paulo and Bangkok, where suspended particulate matter routinely exceeds recommended levels at all the monitoring stations in the city. The effects of emission may be compounded locally by physical geography. Santiago has one of the highest levels of air pollution because the surrounding mountains impede natural ventilation.

Although the general pattern of exchanges is the same in all cities, the magnitudes involved vary widely. Cities do not contribute to environmental damage on an equal basis. Those in developed countries consume disproportionate amounts of resources and, in return, generate excessive amounts of waste. It is estimated that some 25 per cent of the world's population lives in the highly urbanised countries of the developed world, but they consume 70 per cent of the world's energy, 75 per cent of metals and 85 per cent of wood (Tolba and El-Kholy, 1992). The major cities of the developed world draw upon the ecological capital of all other nations to provide food for their economies, and land, air and water in which to discharge their waste products. The Sears Tower in Chicago is a powerful visual symbol of the differential demands of cities since 'this ugly monster uses more energy in twenty-four hours than an average American city of 150,000, or an Indian city of more than one million inhabitants' (Hahn and Simonis, 1991: 12). Research reported by Haughton and Hunter (1994) suggests that urban residents in developed countries generate an average of 0.7–1.8 kg of domestic waste daily, compared to 0.4–0.9 kg daily in developing countries. Measured in per capita terms, urban residents in the USA and Australia are estimated to generate carbon dioxide emissions that are up to 25 times the levels in Dhaka, Bangladesh (Hardoy et al., 1992).

The concern for the sustainability of cities has been expressed at two levels. The first is global and involves a wide range of issues surrounding the long-term stability of the Earth's environment and the implication for cities. It is clear that the world's cities cannot remain prosperous if the aggregate impact of their economies' production and their inhabitants' consumption draws on global resources at unsustainable rates and deposits wastes in global sinks at levels that lead to detrimental climatic change. The second is local and involves the possibility that urban life could be undermined from within because of congestion, pollution and waste generation and their accompanying social and economic consequences. These different concerns focus attention upon the need for intervention at an international scale by governments working together on agreed programmes, and at the domestic level by city authorities addressing the local sustainability issues over which they can exercise some control.

There is growing evidence from climatological research that the Earth's atmosphere is being degraded to an unacceptable extent, with serious implications for life on the planet. There are particular concerns for the well-being of the global climate, which, it is believed, is being threatened by the depletion of the ozone layer and by atmospheric warming (IPCC, 2001). The layer of ozone that exists in the upper atmosphere is of vital importance in the global energy balance because it reduces the amount of harmful solar radiation that is received at the Earth's surface. Ozone occurs when oxygen reacts with ultraviolet light to give a molecule of three oxygen atoms, and its concentration and distribution within the lower stratosphere remain roughly constant under normal conditions. There is growing evidence, however, that the natural cycle of creation and breakdown of ozone in the upper atmosphere has been seriously interrupted by certain compounds, especially chlorofluorocarbons (CFCs). These chemicals are widely used in refrigeration, aerosols, packaging and cleaning, and levels of production have increased significantly in recent years as these applications have grown. CFCs live for a long time and have accumulated in large concentrations in the lower stratosphere, where they are thought to have caused a general thinning of the ozone layer and the appearance of ozone holes. The most extensive hole is that over the Antarctic and there are indications that it is increasing in size and may now cover parts of Australasia and South America. Severe ozone depletion has also been observed during winter months in middle and high latitudes in the northern hemisphere (IPCC, 2001).

It is widely believed that ozone depletion has led to a rise in ultraviolet radiation, which in turn is affecting human health and is threatening many natural and semi-natural ecosystems. The magnitudes are difficult to establish, but a conservative estimate is that a 1 per cent reduction in stratospheric ozone leads to a 3–4 per cent increase in non-melanoma skin cancers (Turner *et al.*, 1994). Sunbathers in areas where the ozone layer is thinnest are at greatest risk. Ozone depletion is also thought to cause eye damage and to suppress people's immune systems. The yield of some commercial food crops and fish stocks may be reduced (Tolba and El-Kholy, 1992).

Pollution is believed to be responsible for a rise in average temperatures across the world, with far-reaching implications for climate, global sea levels and the functioning of local ecosystems. Global temperatures are regulated by a layer of natural 'greenhouse gases' in the atmosphere, including water vapour, carbon dioxide, methane and nitrous oxide. These trap long-wave radiation emitted by the Earth and reflect some of it back to the surface in the form of heat. There is now compelling evidence that a build-up of pollution in the atmosphere has compounded the greenhouse effect and has caused long-term global warming (IPCC, 2001). The principal pollutant is carbon dioxide, which is produced during the burning of fossil fuels, but CFCs are important absorbers of long-wave radiation as well. Atmospheric levels of carbon dioxide have risen by around one-quarter over the past two centuries, with about half of the increase occurring in the last 40 years (Kelly and Karas, 1990). Historically, emissions were higher in the developed world, but the fastest growth today is occurring in developing countries in association with coal-fired heating, inefficient power stations and the rise in the number of motor vehicles.

Although the scientific evidence on the scale of change is equivocal, global warming is widely seen as an increasing long-term threat. The environment is finely balanced and a rise in average global temperatures of as little as 1 per cent could melt polar ice caps sufficiently to raise sea levels across the world by half a metre. Such a change would threaten densely populated areas on deltas and coastal plains. Many major cities, including New Orleans, Amsterdam, Shanghai, Dhaka and Cairo are wholly or partly below present sea level and any rise would significantly increase the risk of flooding. Increases in temperature may also contribute to desertification and reduce agricultural production, especially in areas that are presently semi-arid (McMichael, 1993). Global warming is also predicted to lead to greater climatic fluctuations, accentuating summer temperatures and depressing those in winter. Such large-scale shifts could have particular consequences for cities in semi-arid environments, where climatic conditions are presently marginal.

The threats to cities because of a degradation of the global environment are potentially serious, but they are likely to accrue in the long term. A more immediate possibility is that cities could be seriously undermined from within because of the sheer pressure of numbers on infrastructure and basic services. There are many concerns involved, including fears over the ways in which the built environment is evolving and the implications for the effective functioning of cities as economic and social systems (Haughton and Hunter, 1994). Many cities, especially in the developed world, are suburbanising to such an extent that their coherence and integration is being compromised. Others, particularly in the developing world, are severely stretched to provide public services today, and large numbers of their residents lack basic utilities and amenities, as was emphasised in Chapter 5.

Land availability and use are particular concerns because of fears that cities are in danger of consuming too much of this locally finite resource and developing economic and social forms which are not sustainable. The area of cities increases as populations rise and average household sizes fall and this, combined with greater locational flexibility for individual and industries, has led to urban sprawl in place of the high-density compact urban forms that are associated with the pre-automobile age (Plate 8.1). In some cities the space demands of the car account for almost one-third of the urban land area, rising to two-thirds in inner Los Angeles (McMichael, 1993). Areal expansion is widely seen as inefficient, as it is associated with high energy consumption and increased pollution.

Plate 8.1 *Los Angeles: uncontrolled sprawl threatens the economic and social sustainability of the city*

Evidence is provided by McGlynn *et al.* (1991) who ranked a number of cities from across the world into five groups, from large sprawling US-type cities with high automobile dependence, to compact cities where there is little reliance upon the car. A high positive correlation was found between urban type and environmental impact. Compact cities had fewest adverse consequences for the environment, but sprawling cities had major detrimental effects. Sprawl is further condemned because of its adverse effects on the countryside and because some observers believe that it leads to cities that lack social cohesion, dynamism and vibrancy (Unwin and Searle, 1991). Others point out that, in the long term, urban sprawl is counterproductive. Many of the benefits of the car are short-lived, as rising levels of ownership and use soon lead to congestion and paralysis, undermining the urban structures that the car helped to create.

Although low-density living has many supporters, not least among those who enjoy the environmental attractions of suburbia, there is a widespread view that the physical expansion of cities needs to be checked. There is a limit, probably already exceeded in some countries, to the extent to which the built environment can be allowed to encroach on green land. Such restraint does not necessarily imply a return to compact cities, as high-density living has many disadvantages, including congestion, noise and lack of open space. An alternative way forward is to build new ecological communities based on notions such as permaculture, where the population would be self-reliant and self-sufficient (Orrskog and Snickars, 1992). Such proposals for environmentally efficient settlements are attractive to many, although it is uncertain how they could generate the wealth to support large numbers of people at standards of living that would be acceptable in the twenty-first century. Under present circumstances they seem far-fetched, but they may gain currency if the functioning of cities is seriously undermined by urban sprawl.

Despite the largely pessimistic tone of much of the sustainability literature, there are grounds for believing that the arguments have been overstated. The debate on sustainability is in its infancy and many of the points that are made lack detailed empirical evidence and need to be evaluated in the light of experience. Sceptics are keen to point out that the present concerns are merely the most recent in a string of doomsday predictions for the city, none of which have materialised. Arguments that the population grows more rapidly than food production, thus leading eventually to widespread famine and social breakdown, have been expounded by a succession of writers, from Malthus in the late eighteenth century to Meadows et al. (1972) in their highly influential work on the Limits to Growth. Although differences in growth rates exist, the evidence is that such analysts have consistently underestimated the capacity of farmers to raise outputs. There are famines in several parts of the world today but these are distributional problems – there is no shortage of food overall. The Malthusian paradox is that the agricultural sector in many developed countries is contracting, under strong government pressure, in order to deal with overproduction.

A crisis of food supply is possible at some stage in the future, but at present seems some way off.

A different basis for some optimism about the future is the success of past attempts to deal with urban environmental problems. Most cities in the developed world are cleaner today than they were 20 years ago and their residents enjoy higher levels of health and amenity as a result. Air quality is one area in which improvements have been dramatic. For example, in Greater London, smoke emissions decreased by over 80 per cent in the period 1958–70, despite an increase of at least 30 per cent in output (Beckerman, 1993). The reasons are both technological, involving the switch from coal to oil and gas as energy sources, and political, as governments have introduced and enforced clean air legislation. The drive for clean air continues as the burning of oil and gas create different types of pollutant, but these can now more easily be tackled at source by improving the efficiency of combustion at power stations. The example of improvements in air quality suggest, however, that there are grounds for believing that, with appropriate intervention and direction, a sustainable future for cities is a realistic possibility.

Managing the urban future

The overwhelming message in the sustainability literature is that the urban populations that are envisaged in the first section of this chapter can only be maintained through careful planning and management of resources at both global and local scales. Agreement is required among states to work together with other countries on a common agenda to rehabilitate and to protect the global environment. Action is also necessary to regulate urban development within national boundaries. A start has been made in that most nation states have, under the leadership of the United Nations, entered into commitments to promote sustainable human settlement. The expectation is that governments will set environmental goals for cities and will take appropriate steps to reduce resource use and pollution.

The most comprehensive attempts to date to provide for sustainable development by tackling issues at the global level were the 2002 World Summit on Sustainable Development held in Johannesburg and the 1992 United Nations Conference on Environment and Development held in Rio de Janeiro. The purpose of the Johannesburg Summit was to review progress achieved over the past decade. Some new targets were established or reaffirmed, but the emphasis was on outcomes (Box 8.5). The key focus was on the rate of implementation of the measures that were agreed in Rio for improving environment and development in the twenty-first century (Box 8.6).

Box 8.5 The Johannesburg Summit: key commitments relevant to urban populations

Poverty eradication:

Halve by 2015 the proportion of the world's people whose income is less than US$1 a day, and the proportion who suffer from hunger.

By 2020, achieve a significant improvement in the lives of at least 100 million slum dwellers.

Establish a world solidarity fund to eradicate poverty and to promote social and economic development in developing countries.

Water and sanitation:

Halve, by the year 2015, the proportion of people without access to safe drinking water and to sanitation.

Sustainable production and consumption:

Encourage and promote the development of a ten-year framework of programmes to accelerate the shift towards sustainable consumption and production.

Chemicals:

Aim, by 2020, to use and produce chemicals in ways that do not lead to significant effects upon human health and the environment.

Renew the commitment to the sound management of chemicals and of hazardous wastes throughout their life cycle.

Atmosphere:

Facilitate implementation of the Montreal Protocol on Substances that Deplete the Ozone Layer by ensuring adequate replenishment of its fund by 2003-5.

Improve access by developing countries to alternatives to ozone-depleting substances by 2010, and assist them in complying with the phase-out schedule under the Montreal Protocol.

Health:

Enhance health education with the objective of achieving improved health literacy on a global basis by 2010.

Reduce, by 2015, mortality rates for children under the age of five by two-thirds and maternal mortality rates by three-quarters of the prevailing rate in 2000.

Reduce HIV prevalence among young men and women aged 15–24 by 25 per cent in the most affected countries by 2005 and globally by 2010, as well as combat malaria, tuberculosis and other diseases.

Source: www.johannesburgsummit.org

Box 8.6 The Earth Summit: Agenda 21 programmes for promoting sustainable human settlement development

A Programme area: (A) Providing adequate shelter for all
The objective is to achieve adequate shelter for rapidly growing populations and for the currently deprived urban and rural poor through an enabling approach to shelter development and improvement that is environmentally sound.

B Improving human settlement management
The objective is to ensure sustainable management of all urban settlements, particularly in developing countries, in order to enhance their ability to improve the living conditions of residents, especially the marginalised and disenfranchised, thereby contributing to the achievement of national economic development goals.

C Promoting sustainable land-use planning and management
The objective is to provide for the land requirements of human settlement development through environmentally sound physical planning and land use, so as to ensure access to land for all households and, where appropriate, the encouragement of communally and collectively owned and managed land. Particular attention should be paid to the needs of women and indigenous people for economic and cultural reasons.

D Promoting the integrated provision of environmental infrastructure: water, sanitation, drainage and solid-waste management
The objective is to ensure the provision of adequate environmental infrastructure facilities in all settlements by the year 2025. The achievement of this objective would require that all developing countries incorporate in their national strategies programmes to build the necessary financial and human resource capacity aimed at ensuring better integration of infrastructure and environmental planning by the year 2000.

E Promoting sustainable energy and transport systems in human settlements
The objectives are to extend the provision of more energy-efficient technology and alternative/renewable energy for human settlements and to reduce negative impacts of energy production and use on human health and on the environment.

F Promoting human settlement planning and management in disaster-prone areas
The objective is to enable all countries, especially those which are disaster-prone, to mitigate the negative impact of natural and man-made disasters on human settlements, national economies and the environment.

G Promoting sustainable construction industry activities
The objectives are, first, to adopt policies and technologies and to exchange information on them in order to enable the construction sector to meet human settlement goals, while avoiding harmful side effects on human health and on the biosphere, and, second, to enhance the employment-generation capacity of the construction sector. Governments should work in close collaboration with the private sector in achieving these objectives.

H Promoting human resource development and capacity-building for human settlements development
The objective is to improve human resource development and capacity-building in all countries by enhancing the personal and institutional capacity of all actors, particularly indigenous people and women, involved in human settlement development. In this regard, account should be taken of traditional cultural practices of indigenous people and their relationship to the environment.

Source: Extracted from Johnson (1993: 181–98).

Agenda 21 comprises 40 chapters grouped into four sections. These cover (1) 'social and economic dimensions', (2) 'conservation and management of resources for development', (3) 'strengthening the role of major groups', such as children, women and indigenous peoples, and (4) 'means of implementation'. The seventh chapter of section 1 has the most explicit urban focus, as it outlines an agenda for promoting sustainable human settlement development. It incorporates sets of objectives for eight constituent programme areas (Box 8.6).

The overall human settlement objective of Agenda 21 is to improve the social, economic and environmental quality of human settlements and the living and working conditions of all people, in particular the urban and rural poor. It is envisaged that such improvement will be based upon technical cooperation activities, partnerships among the public, private and community sectors, and participation in the decision-making process by community groups and special interest groups. These approaches should form the core principles of national settlement strategies. In developing these strategies it is anticipated that countries will need to set priorities among the eight programme areas, taking fully into account their social and cultural capabilities.

There was widespread recognition and agreement at the Johannesburg Summit that progress towards implementing sustainable development was extremely disappointing, despite the high ideals and firm intentions that were expressed in Rio. Poverty had deepened and environmental degradation had worsened since 1992. A major reason according to critics, is that the necessary 'global bargain' between the developed and the developing worlds was not struck at Rio (Johnson, 1993). As envisaged, the bargain involves commitments from the developing countries over greenhouse gases, forests and sustainable development, in return for concessions from the developed countries on finance, technology transfer and implementation. For Johnson, the fact that greater agreement was not reached was because of the unwillingness of the developed countries to tackle their profligate lifestyles. The principal culprit is the USA, which has increased its consumption of finite resources and has failed to ratify agreements on atmospheric emissions. This was coupled with the refusal of the developing countries to agree to limit the exploitation of their own natural resources so as to address the imbalance caused by the developed countries that have already used up so much of the Earth's environmental capital and generated so much waste. The absence of agreements on target dates for stabilizing emissions of greenhouse gases and on the protection of forests was another failure contributing to the low level of implementation.

The fact that Earth summits were held at all and that they led to several important agreements and declarations is encouraging, but any benefits will be slow to emerge and will only be measurable in the long term. Particular difficulties surround Agenda 21 since its status is that of a framework for national action. No mandatory rules are specified and there are no clear targets against which progress can be measured. It includes many practical suggestions to assist in achieving sustainable development, but does not have the power or resources to ensure implementation. The governments that attended the summits did not agree to transfer the necessary authority to any international institution. The onus of responsibility is upon individual nations, each of which is likely to respond differently. Few countries have systems of planning and management through which the necessary regulation can be introduced immediately, and many of those that do are committed to minimal intervention in the belief that market mechanisms should decide. Many states have a tradition of acting independently and out of self-interest. The meetings recognised sustainable urban development as an important goal for the next century, but further international agreement and commitment, and concentrated action, are required before rhetoric is likely to be translated into reality.

Conclusion

This book has attempted to analyse and account for the salient characteristics of the urban world and the global city. It has adopted a broad historical and geographical sweep, viewing urban development as a long-term process and seeing the world as being progressively, but as yet incompletely, interlinked and interconnected as an urban place in economic and social terms. Urban growth, urbanisation and the spread of urbanism have yet to run their courses, although many urban changes have happened, especially in the last 30 years, and more are in progress and are likely to occur. 'Urban world' and 'global city' are convenient catchphrases which summarise dominant themes, but they describe conceptual ideals rather than the present situation.

The analysis and discussion in each of the eight chapters has identified key relationships, trends and debates. The global urban population is concentrated in a small number of countries, especially China and India, although the percentage that lives in urban places is high throughout Latin America and the developed world. Within these regions large numbers live in million and mega-cities. Elsewhere, urban populations are smaller and towns and cities are fewer and more widely spaced. Urban places are the predominant form of settlement because they offer significant economies of scale, agglomeration and association. Their emergence reflects the power and persistence of processes that have concentrated large numbers of people in geographically small, yet economically and socially viable, communities. Wide variations, however, exist in the size pattern of cities in different countries, from rank-size distributions at one extreme to primate at the other. Such differences probably reflect the strength of external linkages and, in the case of primate patterns, suggest that the integration of the global urban system is far from complete. A hierarchy of urban places, with world cities at its head, extends across the globe, but separate and localised subsystems that focus upon national primate cities are contained within it.

It is only comparatively recently that the concept of the urban world has begun to have any meaning. Urban development up to the mid twentieth century was largely restricted to the core countries of the world economy. It was most advanced in those parts of north-western Europe and North America which had been industrialised the longest and had dominated extensive political and economic empires. Elsewhere, urban development was embryonic, reflecting the widespread inability of pre-industrial economies to raise productivity and surpluses to levels necessary for significant and sustainable urban growth. When measured at the global scale the wholesale switch of population from rural to urban places is a phenomenon of the last 30 years. It is principally a product of changes in the distribution of population in developing countries. The factors involved are identified and accounted for by interdependency theory, which sees urban growth and urbanisation as consequences of the evolution of capitalism and its changing spatial relations. Of particular importance is the recent globalisation of the economy, a development which is reflected in, and achieved through, the rise of transnational corporations and global finance and producer services institutions, most of which are concentrated in a small number of world cities. The key development is the emergence of a new international division of labour in which production is dispersed to, and so accelerates, urban development in the peripheral areas within the world economy. The effects are reinforced by the commercialisation of agriculture in many developing countries as they engage with the global economy, which has displaced traditional farmers and produced rural to urban migration.

As well as the shift into towns and cities, the world is progressively becoming urbanised in a social and behavioural sense. Traditionally, urban patterns of association and behaviour, although themselves highly varied, were a function of, or related to, place, being restricted to those who actually lived in cities. Today, the lifestyles and values of urbanites are being extended across the globe, both as a direct corollary of urban growth and urbanisation, and because they can be observed, copied and adopted in rural areas via telecommunications and the media. Urban images and messages, once largely western in origin, are becoming more diverse as the producers of media products increase in number. The ability to participate in an urban way of life is increasingly independent of location. The world is fast becoming a global urban society of which we are all residents.

Urban development has contributed to prosperity and success in the core economies, but is associated with disadvantage and deprivation in large parts of the developing world. Many city dwellers in Africa, Asia and South America live in shanty housing, work in informal sector jobs, have limited access to services and suffer poor health. They are trapped in poverty and all its associated consequences for quality of life. Such problems are difficult to resolve because cities are growing most rapidly in the poorest nations, which are presently least able to cope with the social and economic consequences of urban development.

The key issues that surround the urban world concern whether it can continue in its present form. Urban patterns are now well established in most countries, but whether they can absorb further massive increases in urban populations over the next quarter-century seems highly questionable. Doomsday scenarios have been invoked before and have come to nothing, but the sheer scale of likely growth suggests that they must be taken seriously this time. There is ample evidence that the global physical environment is being degraded and that many major cities are near to exhausting their abilities to cope with their exploding populations. Urgent and decisive action is required by governments if a sustainable urban future for all is to be secured.

This book has focused upon wide patterns and has addressed big issues. It has adopted a synoptic approach consistent with the aim of drawing together, overviewing and synthesising the literature, both established and recent, on world urban development. The global perspective and level of analysis will be accepted and endorsed by those who recognise that a primary goal of the social sciences is to pursue and produce general understanding and explanation. It will be appreciated by those who prefer to see big pictures rather than small canvases. Some will question the wisdom of trying to generalise about the distribution, lifestyles and problems of half of the world's population – over three billion people. They will express disappointment and frustration at the lack of intricacy and fine detail without which, they will argue with some justification, the validity of general statements remains untested. The overall purpose, however, is to provide a broad framework within which local theoretical and empirical work can be structured. The need for high-order generalisation is likely to increase as the pace of urban change quickens and as the pattern of global urban development becomes more complex. It took over eight millennia for half the world's population to become urban. Present predictions suggest that it will take less than 80 more years for this process to encompass the remainder.

Recommended reading

Breheny, M. J. (1992) *Sustainable Development and Urban Form*, London: Pion. A useful collection of 14 papers on all aspects of the urban sustainability debate. The principal focus is upon strategic planning for a sustainable future in the developed world. There are valuable case studies from the Netherlands, the UK and Sweden.

Capello, R., Nijkamp, P. and Pepping, G. (1999) *Sustainable Cities and Energy Policies*, Berlin: Springer Verlag. A wide-ranging examination of urban sustainability issues, energy consumption and possible policy responses.

Elliott, J. A. (1999) *An Introduction to Sustainable Development*, London: Routledge. This book evaluates the progress made in the last decades of the twentieth century in achieving sustainable development. There is a concentration upon the developing world and a useful focus upon urban issues.

Hardoy, J. E., Mitlin, D. and Satterthwaite, D. (1992) *Environmental Problems in Third World Cities*, London: Earthscan. A comprehensive description and analysis of the environmental problems of cities in the developing world and how they affect human health, local ecosystems and global cycles. The authors further consider a range of practical solutions and how they could be implemented.

Haughton, G. and Hunter, C. (1994) *Sustainable Cities*, London: Regional Studies Association. This book provides a comprehensive and detailed assessment of the concept of sustainability and how it applies to cities. Key themes are approached from geographical, ecological, economic and managerial perspectives. There is a valuable focus upon policy.

Lee, K., Holland, A. and McNeill, D. (2000) *Global Sustainable Development in the 21st Century*, Edinburgh: Edinburgh University Press. A collection of diverse essays that provide useful background and context to debates on urban sustainability. There is a useful chapter on the meaning of sustainability and the ways in which it is seen by different analysts.

Satterthwaite, D. (1999) *The Earthscan Reader in Sustainable Cities*, London: Earthscan. A collection of 20 papers on all aspects of sustainability and urban development. The treatment is international and there is a strong emphasis on ecological rather than social sustainability.

Key web sites

www.canada.gc.ca   The web site for the Government of Canada. It gives access to the Department of Justice's policy statements on sustainable development.

www.fed.gov.au   The website for the Federal Government of Australia, which gives links to policies on sustainable development.

www.sustainable-development.gov.uk   This is the UK Government's main web site for sustainable development. It provides links to the document *A Better Quality of Life: A Strategy for Sustainable Development*, which outlines the steps taken to deliver on the Agenda 21 undertakings.

www.un.org   This is the main site for the United Nations and provides access, via 'economic and social development', to 'sustainable development', which gives details of the proceedings of conferences, policies and implementation.

www.usinfo.state.gov   The web site for the US State Department. It gives access to the US Government's policy statements on sustainable development.

Topics for discussion

1 What changes do you envisage taking place in the size and geographical distribution of urban populations by 2020? Illustrate your arguments by reference to anticipated developments in a specified country.

2 What do you understand by 'sustainable development'?

3 Critically evaluate the UK Government's approach to sustainable development as set out in *A Better Quality of Life: A Strategy for Sustainable Development*.

4 Critically evaluate the approaches to sustainable urban development of the US Government, the Government of Canada or the Government of Australia.

Appendix: urban sources, definitions and data

Statistics on the urban world are assembled by the Population Division in the Department of Economic and Social Affairs of the United Nations. They are published in its annual *Demographic Yearbook* and biennial *World Urbanisation Prospects* and are accessible on-line at www.un.org. These data form the basis for the urbanisation tables that are included each year in the World Bank's *World Development Reports* and in numerous compilations of world facts and figures and geographical digests. The data are derived from national censuses or, where these are unavailable or inadequate, they are based upon sample surveys or estimates.

The characteristics of the data and the areas to which they refer are of critical importance in comparative urban analysis. Ideally, urban areas should be defined in the same way and full censuses should be taken in each country at the same time on the same basis and with the same known levels of accuracy. The reality is, however, very different, with far-reaching implications for the validity of global urban study.

Definitions

Wide variations exist in the ways in which the populations of countries are divided into urban and rural. There is no standard approach because designation of areas as urban and rural is closely bound up with historical, political, cultural and administrative considerations. A wide range of criteria is used, including size of population, population density, distance between built-up areas, predominant type of economic activity, conformity to legal or administrative status, and characteristics such as specific services and facilities. Urban definitions tend to be revised infrequently, so they soon become outdated. This characteristic has important implications for studies of urban growth and urbanisation.

At first glance, population size would seem to be the most suitable indicator of urban status, but this measure is used in only 26 of the 114 countries and sovereign territories in which census data are available (Table A.1). Even with this simple criterion, wide variations exist as to the required minima. Places with as few as 200 inhabitants are considered to be urban in Iceland, while in Switzerland and Malaysia the lower limit is 10,000 people. Important differences exist between countries over who to include in the urban population. Some censuses record all those who are present on a particular date, whereas others attempt to enumerate those who are normally resident. The latter results in grossly inflated figures, where there are large numbers of people who are only nominally resident in the city but who normally work and live away. This is the case in many developing countries, where there are many migrant workers. It is especially serious in China, where the city in which people are officially registered as living is often different to that in which they actually reside.

In 20 countries, population size is combined with other diagnostic criteria. Examples include Israel, Botswana and Zambia, which use population size and employment in non-agricultural occupations; Canada and France, which employ size and density criteria; and India, where consideration is given to population, density, legal and morphological characteristics. Such indices provide a more satisfactory basis for urban definition within the country concerned, although the composition and threshold values used again mean that the difficulty of making meaningful cross-national comparisons of urban patterns and problems is increased.

Urban places in the vast majority of countries are identified on a legal, administrative or governmental basis. No attempt is made to establish objective criteria. Instead, urban status is deemed to apply to all those places that occupy a particular level within the municipal hierarchy. This practice is especially common in small countries, where towns and cities are designated by government decree. Thus, the urban places in Costa Rica are the administrative centres of cantons; in Pakistan they are places with municipal corporation, town committee or cantonment; in South Africa they are places with some form of local authority; and in Iraq they are areas within the boundaries of municipality councils.

Global urban study is even less well served by 15 of the world's national governments from which urban criteria are 'not available'. The bases for urban definition, and how these compare with those in other countries, are therefore unknown. This group includes Colombia, Myanmar, Philippines, Somalia, Uganda and Belize.

Table A.1 *Definitions of urban places: selected examples*

Criterion	Number	Examples
Population	26	Ethiopia: localities of 2,000 or more inhabitants.
		Venezuela: centres with a population of 1,000 or more inhabitants.
		Malaysia: gazetted areas with a population of 10,000 or more.
		Albania: towns and other industrial centres of more than 400 inhabitants.
		Iceland: localities of 200 or more inhabitants.
		Switzerland: communes of 10,000 or more inhabitants, including suburbs.
		Ireland: cities and towns, including suburbs, of 1,500 or more inhabitants.
Population plus additional criteria	20	Zambia: localities of 5,000 or more inhabitants, the majority of whom depend on non-agricultural activities.
		Botswana: agglomerations of 5,000 or more inhabitants where 75 per cent of the economic activity is of the non-agricultural type.
		Canada: places of 1,000 or more inhabitants having a population density of 400 or more per square kilometre.
		Israel: all settlements of more than 2,000 inhabitants, except those where at least one-third of households, participating in the civilian labour force, earn their living from agriculture.
		France: communes containing an agglomeration of more than 2,000 inhabitants living in contiguous houses with not more than 200 metres between houses; also, communes of which the major portion of the population is part of a multicultural agglomeration of this nature.
		India: towns (places with municipal corporation, municipal area committee, towns committee, notified area committee or cantonment board); also, all places having 5,000 or more inhabitants, a density of not less than 1,000 people per square mile or 390 per square kilometre, pronounced urban characteristics and at least three-quarters of the adult male population employed in pursuits other than agriculture.
Legal, administrative governmental	54	Bangladesh: places have a municipality (*pourashava*), a town committee, *shahar* committee or a cantonment board.

Table A.1 *continued*

Criterion	Number	Examples
Legal, administrative governmental	54	Pakistan: places with municipal corporation, town committee or cantonment.
		South Africa: places with some form of local authority.
		Swaziland: localities proclaimed as urban.
		Iraq: the area within the boundaries of Municipality Councils.
		Mongolia: capital and district centres.
		Brazil: urban and suburban zones of administrative centres of *municipos* and districts.
		Uruguay: cities.
Not available	14	Myanmar
		Philippines
		Colombia
Total	114	

Source: United Nations (2002: chapter 8).

Data

As well as objectivity and consistency of definition, comparative international study is further impaired by variations in the frequency and timing of national population censuses. Following the lead established in Britain in 1801, most developed countries undertake a census of population, both urban and rural, every ten years. Censuses are held with due regard to the principles of social surveying and yield results with a known degree of statistical reliability. They form the basis for annual estimates of the size of the urban population. The reliability of such estimates is related to the time elapsed since the last census.

The practice elsewhere in the world is, however, more varied. Some countries have never held a census of their populations. In others, censuses are an occasional and infrequent rather than regular occurrence. For example, the last census to be held in Angola, Djibouti and Togo was in 1970. According to the United Nations, there are 20 major countries that have not held a national census since 1985 (Table A.2). The list includes many of the poorest countries of Africa and Asia.

Table A.2 *Major countries which have not held a national population census since 1985*

Country	Most recent census
Africa	
Angola	1970
Congo	1984
Democratic Republic of Congo	1984
Djibouti	1970
Guinea-Bissau	1979
Kenya	1979
Liberia	1984
Libya	1984
Togo	1970
Central America	
Guatemala	1981
Asia	
Afghanistan	1979
Lebanon	1980
Myanmar	1983
Saudi Arabia	1974
United Arab Emirates	1980
Europe	
Denmark	1981
Germany	1981
Luxembourg	1970
Netherlands	1960
Russia	1989

Source: United Nations (2002: Chapter 8).

Data on the present day urban population of such countries are generated through extrapolation. Their reliability is especially low.

Variations in urban definition and in the availability and quality of urban data introduce significant but unknown biases into comparative urban study. They are of particular importance in analysing contemporary urban change because the quality of definitions and data are commonly lowest in precisely those countries in which the urban population is largest and is growing the quickest. International standardisation of urban definitions and methods of data collection is a remote prospect. Global urban data are best regarded as crude estimates which support only general rather than precise statements about the distribution and growth of population in the contemporary urban world.

Recommended reading

United Nations (2002) *World Urbanisation Prospects: The 2001 Revision*, New York: United Nations. Chapter 8 of this publication considers the problem of urban definition and specifies sources of urban data for each country.

Key web site

www.un.org   This site provides access, via 'economic and social development' and 'population', to urban data in the biennial reports on *World Urbanisation Prospects*.

References

Abu-Lughod, J. L. (1995) Comparing Chicago, New York and Los Angeles: testing some world city hypotheses, in Knox, P. L. and Taylor, P. J. (eds) *World Cities in a World System*, Cambridge: Cambridge University Press, 171–91.

Adams, R. M. (1966) *The Evolution of Urban Society*, London: Weidenfeld and Nicolson.

Akwule, R. (1992) *Global Telecommunications: Technology, Administration and Policies*, London: Focal Press.

Alcamo, J. and Lubkert, B. (1990) The city and the air: Europe, in Canfield, C. (ed.) *Ecocity Conference, 1990*, Berkeley: Urban Ecology, 12–20.

Alvarado, M. (1989) *Global Video*, London: UNESCO.

Appadurai, A. (1996) *Modernity at Large: Cultural Dimensions of Globalization*, Minneapolis: University of Minneapolis Press.

Auty, R. (1995) *Patterns of Development: Resources, Policy and Economic Growth*, London: Arnold.

Barrett, H. R. and Browne, A. W. (1996) Export horticultural production in sub-Saharan Africa: the incorporation of The Gambia, *Geography* 81, 47–56.

Barrett, H. R., Ilbery, B. W., Browne, A. W. and Binns, T. (1999) Globalisation and the changing networks of food supply: the importation of fresh agricultural produce from Kenya into the UK, *Transactions of the Institute of British Geographers* 24, 159–74.

Barrios, L. (1988) Television, telenovelas and family life in Venezuela, in Lull, J. (ed.) *World Families Watch Television*, London: Sage, 49–79.

Beaverstock, J. V., Smith R. G. and Taylor P. J. (1999) A roster of world cities, *Cities* 16, 445–58.

Beavon, K. (1977) *Central Place Theory: A Reinterpretation*, London: Longman.

Beckerman, W. (1993) The environmental limits to growth: a fresh look, in Giersch, H. (ed.) *Economic Progress and Environmental Concern*, Berlin: Springer Verlag, 32–48.

Bekkers, W. (1987) The Dutch public broadcasting services in a multi-channel environment, in ESOMAR (eds) *The Application of Research to Broadcasting Decisions*, London: European Society of Market Research, 169–88.

Berg, L. van den, Drewett, R., Klassen, L. H., Rossi, A. and Vijverberg, C. H. T. (1982) *A Study of Growth and Decline*, London: Pergamon.

Berry, B. J. L. (1961) City size distribution and economic development, *Economic Development and Cultural Change* 9, 573–87.

Berry, B. J. L. (1976) *Urbanisation and Counterurbanisation*, London: Sage.

Blowers, A. (1993) *Planning for a Sustainable Environment*, London: Earthscan.

Bowen, D. (1986) The class of 86, *Business* November, 34–41.

Bradnock, W. (1984) *Urbanisation in India*, London: Murray.

Breeze, G. (1972) *The City in Newly Developing Countries: Readings in Urbanism and Urbanisation*, London: Prentice Hall.

Brunn, S. D. and Williams, J. F. (1993) *Cities of the World: World Regional Urban Development*, New York: HarperCollins.

Camigni, R., Capello, R. and Nijkamp, P. (2001) Managing sustainable urban environments, in Paddison, R. (ed.) *Handbook of Urban Studies*, London: Sage, 124–39.

Capello, R., Nijkamp, P. and Pepping, G. (1999) *Sustainable Cities and Energy Policies*, Berlin: Springer Verlag.

Carter, H. and Lewis, C. R. (1991) *An Urban Geography of England and Wales in the Nineteenth Century*, London: Arnold.

Castells, M. (1977) *The Urban Question: A Marxist Approach*, London: Edward Arnold.

Castells, M. (1989) *The Informational City: Information Technology, Economic Restructuring and the Urban-Regional Process*, Oxford: Blackwell.

Castells, M. (1996) *The Rise of Network Society*, Oxford: Blackwell.

Champion, A. G. (1989) *Counterurbanisation*, London: Arnold.

Champion, A. G. (1999) Urbanisation and counterurbanisation, in Pacione, M. (ed.) *Applied Geography: Principles and Practice*, London: Routledge, 347–57.

Champion, A. G. (2001) Urbanisation, suburbanisation, counterurbanisation and reurbanisation, in Paddison, R. (ed.) *Handbook of Urban Studies*, London: Sage, 143–61.

Chase-Dunn, C. (1985) The system of world cities: AD 800–1975, in Timberlake, M. (ed.) *Urbanisation in the World Economy*, New York: Academic Press, 269–92.

Chase-Dunn, C. (1989) *Global Formation: Structures of the World Economy*, London: Blackwell.

Childe, V. G. (1950) The urban revolution, *Town Planning Review* 21, 3–17.

Christaller, W. (1933) *Die Zentralen Orte in Suddeutschland*, Jena: Gustav Fischer, translated by Baskin, C. W. (1966) as *Central Places in Southern Germany*, Englewood Cliffs, NJ: Prentice Hall.

Clark, D. (1989) *Urban Decline*, London: Routledge.

Clark, D. (1998) Interdependent urbanisation in an urban world: an historical overview, *The Geographical Journal* 164, 85–95.

Clarke, W. M. (2001) *How the City of London Works*, London: Sweet and Maxwell.

Cochrane, A. (1995) Global worlds and worlds of difference, in Anderson, J., Brook, C. and Cochrane, A. (eds) *A Global World*, Oxford: Oxford University Press, 249–80.

Cross, D. F. W. (1990) *Counterurbanisation in England and Wales*, London: Avebury.

Cubitt, T. (1995) *Latin American Society*, London: Longman.

Cumming, S. D. (1990) Post-colonial urban residential change in Zimbabwe: a case study, in Potter, R. B. and Salau, A. T. (eds) *Cities and Development in the Third World*, London: Mansell, 32–50.

Davis, K. (1965) The urbanisation of the human population, *Scientific American* 213, 40–53.

Davis, K. (1969) *World Urbanisation*, Los Angeles: University of California.

DES and the Welsh Office (1990) *Geography for Ages 5 to 16*, London: Department of Education and Science and The Welsh Office.

de Soto, H. (1989) *The Other Path: The Invisible Revolution in the Third World*, New York: Harper and Row.

Dicken, P. (1998) *Global Shift*, London: Harper and Row.

Drakakis-Smith, D. (1992) *Urban and Regional Change in Southern Africa*, London: Routledge.

El Shakhs, S. (1972) Development, primacy and systems of cities, *The Journal of Developing Areas* 7 (October), 11–36.

Fainstein, S. S., Gordon, I. and Harloe, M. (eds) (1992) *Divided Cities: New York and London in the Contemporary World*, Oxford: Blackwell.

Fielding, A. J. (1989) Counterurbanization in Western Europe, *Progress in Planning* 17, 1–52.

Findlay, A. M., Li, F. L. N., Jowett, A. J. and Skeldon, R. (1996) Skilled international migration and the global city: a study of expatriates in Hong Kong, *Transactions of the Institute of British Geographers* 21, 49–61.

Findley, S. E. (1993) The Third World city: development policy and issues, in Kasarda, J. D. and Parnell, A. M. (eds) *Third World Cities*, London: Sage, 1–33.

Forbes, D. and Thrift, N. J. (1987) International impacts on the urbanisation process in the Asian region: a review, in Fuchs, R. J., Jones, G. W. and Pernia, E. M. (eds) *Urbanisation and Urban Policies in Pacific Asia*, Boulder, CO: Westview.

Frank, A. G. (1967) *Capitalism and Underdevelopment in Latin America: Historical Studies of Chile and Brazil*, New York: Monthly Review Press.

Frank, A. G. (1969) *Latin America: Underdevelopment or Revolution?*, New York: Monthly Review Press.

Friedmann, H. (1993) The political economy of food, *New Left Review* 197, 29–57.

Friedmann, J. (1972) The spatial organisation of power in the development of urban systems, *Development and Change* 4, 12–50.

Friedmann, J. (1986) The world city hypothesis, *Development and Change* 17, 69–74.

Frobel, F., Heinrichs, J. and Kreye, O. (1980) *The New International Division of Labour*, Cambridge: Cambridge University Press.

Fujita, K. (1991) A world city and flexible specialization: restructuring of the Tokyo metropolis, *International Journal of Urban and Regional Research* 15, 269–84.

Giddens, A. (1990) *The Consequences of Modernity*, California: Stanford University Press.

Gilbert, A. and Gugler, J. (1992) *Cities, Poverty and Development*, Oxford: Oxford University Press.

Giradet, H. (1990) The metabolism of cities, in Cadman, D. and Payne, G. (eds) *The Living City*, London: Routledge, 170–80.

Giradet, H. (1998) Sustainable cities: a contradiction in terms?, *Journal of the Scottish Association of Geography Teachers* 27, 50–7.

Goldstein, S. (1989) Levels of urbanisation in China, in Dogan, M. and Kasarda, J. D. (eds) *The Metropolis Era, Vol. 1: A World of Great Cities*, London: Sage, 187–225.

Goldstein, S. (1993) The impact of temporary migration on urban places: Thailand and China as case studies, in Kasarda, J. D. and Parnell, A. M. (eds) *Third World Cities*, London: Sage, 199–219.

Goodman, D. and Watts, M. (1997) *Globalizing Food*, London: Routledge.

Goudie, A. (1990) *The Human Impact on the Natural Environment*, Oxford: Blackwell.

Griffin, E. and Ford, L. (1993) Cities of Latin America, in Brunn, S. D. and Williams, J. F. (eds) *Cities of the World*, New York: HarperCollins, 225–66.

Gugler, J. (1988) *The Urbanisation of the Third World,* Oxford: Oxford University Press.

Hahn, E. and Simonis, U. (1990) Ecological urban restructuring: method and action, *Environmental Management and Health* 2, 12–19.

Haider, D. (1992) Place wars: new realities of the 1990s, *Economic Development Quarterly* 6, 127–34.

Hall, P. (1966) (3rd edn 1984) *The World Cities*, London: Weidenfeld and Nicolson.

Hamnett, C. (1994) Social polarisation in global cities: theory and evidence, *Urban Studies* 31, 401–24.

Hamnett, C. (1996) Why Sassen is wrong: a response to Burgers, *Urban Studies* 33, 107–10.

Hannerz, U. (1997) Scenarios for peripheral cultures, in King, A. D. (ed.) *Culture, Globalisation and the World-System: Contemporary Conditions for the Representation of Identity*, Minneapolis: University of Minnesota Press, 107–28.

Hardoy, J. E. and Satterthwaite, D. (1989) *Squatter Citizen*, London: Earthscan.

Hardoy, J. E. and Satterthwaite, D. (1990) Urban change in the Third World: are recent trends a useful pointer to the urban future?, in Cadman, D. and Payne, G. (eds) *The Living City*, London: Routledge, 75–110.

Hardoy, J. E., Mitlin, D. and Satterthwaite, D. (1992) *Environmental Problems in Third World Cities*, London: Earthscan.

Harvey, D. W. (1973) *Social Justice and the City*, London: Edward Arnold.

Haughton, G. and Hunter, C. (1994) *Sustainable Cities*, London: Regional Studies Association.

Ilbery, B. W. (2001) Changing geographies of global food production, in Daniels, P., Bradshaw, M., Shaw, D. and Sidaway, J. (eds) *Human Geography: Issues for the Twenty-first Century*, Harlow: Pearson, 253–73.

ILO (1972) *Employment, Incomes and Equality: A Strategy for Increasing Productive Employment in Kenya*, Geneva: International Labour Organisation.

IPCC (2001) *Climatic Change 2001: the Scientific Evidence*, Cambridge: Cambridge University Press.

Johnson, S. P. (1993) *The Earth Summit: The United Nations Conference on Environment and Development*, London: Graham and Trotman.

Johnston, R. J. (1980) *City and Society: An Outline for Urban Geography*, Harmondsworth: Penguin.

Johnston, R. J., Taylor, P. J. and Watts, M. J. (2002) *Geographies of Global Change: Remapping the World*, Oxford: Blackwell.

Kasarda, J. D. and Parnell, A. M. (1993) *Third World Cities*, London: Sage.

Kelly, P. M. and Karas, J. H. W. (1990) The greenhouse effect, *Capital and Class* 38, 17–27.

Khan, H. and Weiner, A. J. (1967) *The Year 2000*, New York: Macmillan.

King, A. D. (1989) *Global Cities: Post-imperialism and the Internationalisation of London*, London: Routledge.

King, A. D. (1990) *Urbanism, Colonialism and the World Economy: Cultural and Spatial Foundations of the World Urban System*, London: Routledge.

Knight, R. V. and Gappert, G. (1989) *Cities in a Global Society*, New York: Sage.

Knox, P. and Agnew, J. (1994) *The Geography of the World Economy*, London: Arnold.

Lampard, E. E. (1955) The history of cities in economically advanced areas, *Economic Development and Cultural Change* 3, 81–102.

Lampard, E. E. (1965) Historical aspects of urbanisation, in Hauser, P. M. and Schnore, L. F. (eds) *The Study of Urbanisation*, London: Wiley, 519–54.

Leyshon, A. and Thrift, N. (1997) *Money Space: Geographies of Monetary Transformation*, London: Routledge.

Lull, J. (1995) *Media, Communication and Culture: A Global Approach*, Cambridge: Polity Press.

Lull, J. and Se-Wen, S. (1988) Agents of modernisation: television and urban Chinese families, in Lull, J. (ed.) *World Families Watch Television*, London: Sage, 193–236.

Machimura, T. (1992) The urban restructuring process in the 1980s: transforming Tokyo into a world city, *International Journal of Urban and Regional Research* 16, 114–29.

Machimura, Y. (1998) Symbolic use of globalization in urban politics in Tokyo, *International Journal of Urban and Regional Research* 22, 183–94.

Markusen, A. and Gwiasda, V. (1994) Multipolarity and the layering of functions in world cities: New York City's struggle to stay on top, *International Journal of Urban and Regional Research* 18, 167–93.

Mattelart, A. (1979) *Multinational Corporations and the Control of Culture*, London: Harvester Press.

McEwan, C. (2001) Geography, culture and global change, in Daniels, P., Bradshaw, M., Shaw, D. and Sidaway, J. (eds) *Human Geography: Issues for the Twenty-first Century*, Harlow: Pearson, 154–80.

McGee, T. G. (1967) *The Southeast Asian City*, London: Bell.

McGlynn, G., Newman, P. and Kenworthy, J. (1991) Land use and transport: the missing link in urban consolidation, *Urban Futures* special issue 1 (July), 8–18.

McLuhan, H. M. (1964) *Understanding Media: The Extensions of Man*, New York: McGraw-Hill.

McMichael, M. (1993) *Planetary Overload: Global Environmental Change and the Health of the Human Species*, Cambridge: Cambridge University Press.

Meadows, D. H., Meadows, D. L., Randers, J. and Behrenv, W. W. (1972) *The Limits to Growth*, New York: University Books.

Meier, R. L. (1962) *A Communications Theory of Urban Growth*, Cambridge, MA: Massachusetts Institute of Technology Press.

Mitchell, R. D. (2001) The colonial origins of Anglo-America, in McIlwraith, T. F. and Muller, E. K. (eds) *North America: The Historical Geography of a Changing Continent*, New York: Rowman and Littlefield, 89–118.

Morley, D. and Robins, K. (1995) *Spaces of Identity: Global Media, Electronic Landscapes and Cultural Boundaries*, London: Routledge.

O'Connor, J. (1998) Popular culture, cultural intermediaries and urban regeneration, in Hall, T. and Hubbard, P. (eds) *The Entrepreneurial City: Geographies of Politics, Regime and Representation*, Chichester: Wiley.

Orrskog, L. and Snickars, F. (1992) On the sustainability of urban and regional structures, in Breheny, M. (ed.) *Sustainable Development and Urban Form*, London: Pion.

Pacione, M. (2001) *Urban Geography: A Global Perspective*, London: Routledge.

Patel, S. (1985) Street children, hotel boys and children of pavement dwellers in Bombay: how they meet their daily needs, *Environment and Urbanisation* 2, 9–26.

Patterson, R. (1987) *International TV and Video Guide 1987*, London: Tantivy Press.

Pearce, D., Markandya, A. and Barbier, E. B. (1989) *Blueprint for a Green Economy*, London: Earthscan.

Pred, A. R. (1977) *City Systems in Advanced Economies*, London: Hutchinson.

Preston, S. H. (1988) Urban growth in developing countries: a demographic reappraisal, in Gugler, J. (ed.) *The Urbanisation of the Third World*, Oxford: Oxford University Press, 11–32.

Ramachandaran, R. (1993) *Urbanisation and Urban Systems in India*, Oxford: Oxford University Press.

Rondinelli, D. (1989) Giant and secondary city growth in Africa, in Dogan, M. and Kasarda, J. D. (eds) *The Metropolis Era, Vol. 1: A World of Great Cities*, London: Sage, 291–321.

Sassen, S. (1991) *The Global City: New York, London, Tokyo*, Princeton, NJ: Princeton University Press.

Sassen, S. (1994) *Cities in a World Economy*, London: Pine Forge.

Schiller, H. (1976) *Communication and Cultural Domination*, White Plains, NY: International Arts and Sciences Press.

Short, J. R. (1996) *The Urban Order: An Introduction to Cities, Culture and Power*, Cambridge: Blackwell.

Short, J. R. and Kim, Y.-H. (1999) *Globalization and the City*, Harlow: Pearson.

Sinclair, J., Jacka, E. and Cunningham, S. (eds) (1996) *New Patterns in Global Television: Peripheral Vision*, Oxford: Oxford University Press.

Sit, V. F. S. (1985) *Chinese Cities: The Growth of the Metropolis Since 1949*, Oxford: Oxford University Press.

Sit, V. F. S. (1993) Transnational capital flows, foreign investments and urban growth in developing countries, in Kasarda, J. D. and Parnell, A. M. (eds) *Third World Cities*, London: Sage, 180–98.

Slater, T. (2001) The rise and spread of capitalism, in Daniels, P., Bradshaw, M., Shaw, D. and Sidaway, J. (eds) *Human Geography: Issues for the 21st Century*, Harlow: Pearson, 41–72.

Smith, C. A. (1985a) Theories and measures of urban primacy: a critique, in Timberlake, M. (ed.) *Urbanisation in the World-Economy*, London: Academic Press, 87–116.

Smith, C. A. (1985b) Class relations and urbanisation in Guatemala: towards an alternative theory of urban primacy, in Timberlake, M. (ed.) *Urbanisation in the World Economy*, London: Academic Press, 121–59.

Sreberny-Mohammadi, A. (1991) The global and the local in international communications, in Curran, J. and Gurevitch, M. (eds) *Mass Media and Society*, London: Arnold, 118–38.

Stewart, D. J. (1997) African urbanization: dependent linkages in a global economy, *Tidjschrift voor Economische en Social Geografie* 88, 251–61.

Stoneman, C. (1979) Foreign capital and the reconstruction of Zimbabwe, *Review of African Political Economy* 11, 62–83.

Taaffe, E. J., Morrill, R. L. and Gould, P. R. (1963) Transport expansion in underdeveloped countries: a comparative analysis, *Geographical Review* 53, 503–29.

Taylor, P. J. (1993) *Political Geography: World Economy, Nation-State and Locality*, London: Longman.

Thomas, J. J. (1995) *Surviving the City*, London: Pluto.

Thorngren, B. (1970) How do contact systems affect regional development?, *Environment and Planning* 2, 409–27.

Thrift, N. (2002) A hyperactive world, in Johnston, R. J., Taylor, P. J. and Watts, M. J. (eds) *Geographies of Global Change*, Oxford: Blackwell, 29–42.

Tolba, M. K. and El-Kholy, O. A. (1992) *The World Environment, 1972–92: Two Decades of Challenge*, London: Chapman and Hall.

Torado, M. P. (2000) *Economic Development*, London: Longman.

Tracey, M. (1993) A taste of money: popular culture and the economics of global television, in Noam, E. M. and Millonzi, J. C. (eds) *The International Market in Film and Television Programs*, Norwood, NJ: Ablex, 163–98.

Turner, R. K., Pearce, D. and Bateman, I. (1994) *Environmental Economics: An Elementary Introduction*, Hemel Hempstead: Harvester Wheatsheaf.

Ullman, E. L. and Dacey, M. F. (1962) The minimum requirements approach to the urban economic base, in Norborg, K. (ed.) *Proceedings of the I.G.U. Symposium on Urban Geography*, Lund: C. W. K. Gleerup, 485–518.

UNCTAD (2002) *World Investment Report 2002: Transnational Corporations and Export Competitiveness*, New York: United Nations.

UNDP (2002) *Human Development Report*, New York: United Nations Development Programme.

UNESCO (1999) *World Information and Communication Report, 1999–2002*, Paris: UNESCO.

United Nations (2001a) *Cities in a Globalizing World: Global Report on Human Settlements*, London: Earthscan.

United Nations (2001b) *The State of the World's Cities*, New York: United Nations.

United Nations (2002) *World Urbanisation Prospects*, New York: United Nations.

United Nations (2003) *Demographic Yearbook*, New York: United Nations.

Unwin, N. and Searle, G. (1991) Ecologically sustainable development and urban development, *Urban Futures* special issue 4 (November), 1–12.

Vance, J. E. (1970) *The Merchant's World: The Geography of Wholesaling*, Englewood Cliffs, NJ: Prentice Hall.

Wallerstein, I. (1979) *The Capitalist World Economy*, New York: Cambridge University Press.

Wallerstein, I. (1980) *The Modern World-System 2: Mercantilism and the Consolidation of the European World Economy, 1600–1750*, London: Academic Press.

Wallerstein, I. (1989) *The Modern World-System 3: The Second Era of the Great Expansion of the Capitalist World Economy, 1730–1840s*, London: Academic Press.

WCED (1987) *Our Common Future*, Oxford: Oxford University Press.

Webber, M. M. (1964) The urban place and the non-place urban realm, in Webber, M. M., Dyckham, J. W., Foley, D. L., Guttenberg, A. Z., Wheaton,

W. L. C. and Wurster, C. B. (eds) *Explorations into Urban Structure* Philadelphia: University of Pennsylvania Press, 79–153.

Weber, A. F. (1899) *The Growth of Cities in the Nineteenth Century*, New York: Macmillan. 1962 reprint, Ithaca, NY: Cornell University Press.

Wheatley, P. (1971) *The Pivot of the Four Quarters*, Chicago: The University of Chicago Press.

WHO (2002) *World Health Report 2002*, Geneva: World Health Organization.

Wilhelmy, H. (1986) Urban change in Argentina: historical roots and modern trends, in Conzen, M. P. (ed.) *World Patterns of Modern Urban Change*, Chicago: University of Chicago Research Paper 217–18, 273–329.

Williams, S. (1992) The coming of the groundscapers, in Budd, L. and Whimster, S. (eds) *Global Finance and Urban Living*, London: Routledge, 246–59.

Wirth, L. (1938) Urbanism as a way of life, *American Journal of Sociology* 44, 1–24.

World Bank (2002) *World Development Report, 2002*, New York: World Bank.

Zipf, G. K. (1949) *Human Behavior and the Principle of Least Effort*, New York: Addison and Wesley.

Zukin, S. (1992) The city as a landscape of power: London and New York as global financial capitals, in Budd, L. and Whimster, S. (eds) *Global Finance and Urban Living*, London: Routledge, 195–233.

Index